Shared-Memory Synchronization

Synthesis Lectures on Computer Architecture

Editor
Mark D. Hill, *University of Wisconsin, Madison*

Synthesis Lectures on Computer Architecture publishes 50- to 100-page publications on topics pertaining to the science and art of designing, analyzing, selecting and interconnecting hardware components to create computers that meet functional, performance and cost goals. The scope will largely follow the purview of premier computer architecture conferences, such as ISCA, HPCA, MICRO, and ASPLOS.

A Primer on Memory Consistency and Cache Coherence
Daniel J. Sorin, Mark D. Hill, and David A. Wood
2011

Dynamic Binary Modification: Tools, Techniques, and Applications
Kim Hazelwood
2011

Quantum Computing for Computer Architects, Second Edition
Tzvetan S. Metodi, Arvin I. Faruque, and Frederic T. Chong
2011

High Performance Datacenter Networks: Architectures, Algorithms, and Opportunities
Dennis Abts and John Kim
2011

Processor Microarchitecture: An Implementation Perspective
Antonio González, Fernando Latorre, and Grigorios Magklis
2010

Transactional Memory, 2nd edition
Tim Harris, James Larus, and Ravi Rajwar
2010

Computer Architecture Performance Evaluation Methods
Lieven Eeckhout
2010

Introduction to Reconfigurable Supercomputing
Marco Lanzagorta, Stephen Bique, and Robert Rosenberg
2009

On-Chip Networks
Natalie Enright Jerger and Li-Shiuan Peh
2009

The Memory System: You Can't Avoid It, You Can't Ignore It, You Can't Fake It
Bruce Jacob
2009

Fault Tolerant Computer Architecture
Daniel J. Sorin
2009

The Datacenter as a Computer: An Introduction to the Design of Warehouse-Scale Machines
Luiz André Barroso and Urs Hölzle
2009

Computer Architecture Techniques for Power-Efficiency
Stefanos Kaxiras and Margaret Martonosi
2008

Chip Multiprocessor Architecture: Techniques to Improve Throughput and Latency
Kunle Olukotun, Lance Hammond, and James Laudon
2007

Transactional Memory
James R. Larus and Ravi Rajwar
2006

Quantum Computing for Computer Architects
Tzvetan S. Metodi and Frederic T. Chong
2006

Shared-Memory Synchronization

Michael L. Scott

ISBN: 978-3-031-00612-8   paperback
ISBN: 978-3-031-01740-7   ebook

DOI 10.1007/978-3-031-01740-7

A Publication in the Springer series
SYNTHESIS LECTURES ON ADVANCES IN AUTOMOTIVE TECHNOLOGY

Lecture #23
Series Editor: Mark D. Hill, *University of Wisconsin, Madison*
Series ISSN
Synthesis Lectures on Computer Architecture
Print 1935-3235   Electronic 1935-3243

Shared-Memory Synchronization

Michael L. Scott
University of Rochester

SYNTHESIS LECTURES ON COMPUTER ARCHITECTURE #23

ABSTRACT

Since the advent of time sharing in the 1960s, designers of concurrent and parallel systems have needed to synchronize the activities of threads of control that share data structures in memory. In recent years, the study of synchronization has gained new urgency with the proliferation of multicore processors, on which even relatively simple user-level programs must frequently run in parallel.

This lecture offers a comprehensive survey of shared-memory synchronization, with an emphasis on "systems-level" issues. It includes sufficient coverage of architectural details to understand correctness and performance on modern multicore machines, and sufficient coverage of higher-level issues to understand how synchronization is embedded in modern programming languages.

The primary intended audience is "systems programmers"—the authors of operating systems, library packages, language run-time systems, concurrent data structures, and server and utility programs. Much of the discussion should also be of interest to application programmers who want to make good use of the synchronization mechanisms available to them, and to computer architects who want to understand the ramifications of their design decisions on systems-level code.

KEYWORDS

atomicity, barriers, busy-waiting, conditions, locality, locking, memory models, monitors, multiprocessor architecture, nonblocking algorithms, scheduling, semaphores, synchronization, transactional memory

*To Kelly, my wife and partner
of more than 30 years.*

Contents

Preface

This lecture grows out of some 25 years of experience in synchronization and concurrent data structures. Though written primarily from the perspective of systems software, it reflects my conviction that the field cannot be understood without a solid grounding in both concurrency theory and computer architecture.

Chapters 4, 5, and 7 are in some sense the heart of the lecture: they cover spin locks, busy-wait condition synchronization (barriers in particular), and scheduler-based synchronization, respectively. To set the stage for these, Chapter 2 surveys aspects of multicore and multiprocessor architecture that significantly impact the design or performance of synchronizing code, and Chapter 3 introduces formal concepts that illuminate issues of feasibility and correctness.

Chapter 6 considers atomicity mechanisms that have been optimized for the important special case in which most operations are read-only. Later, Chapter 8 provides a brief introduction to *nonblocking algorithms*, which are designed in such a way that all possible thread interleavings are correct. Chapter 9 provides a similarly brief introduction to *transactional memory*, which uses speculation to implement atomicity without (in typical cases) requiring mutual exclusion. (A full treatment of both of these topics is beyond the scope of the lecture.)

Given the volume of material, readers with limited time may wish to sample topics of particular interest. All readers, however, should make sure they are familiar with the material in Chapters 1 through 3. In my experience, practitioners often underestimate the value of formal foundations, and theoreticians are sometimes vague about the nature and impact of architectural constraints. Readers may also wish to bookmark Table 2.1 (page 19), which describes the memory model assumed by the pseudocode. Beyond that:

- Computer architects interested in the systems implications of modern multicore hardware may wish to focus on Sections 2.2–2.3.1, 3.3–3.4, 4.2–4.3, 4.5.1, 5.1–5.2, 8.1–8.3, and 9.2.

- Programmers with an interest in operating systems and run-time packages may wish to focus on Sections 2.2–2.3.1, all of Chapters 3–6, and Section 7.5.

- Authors of parallel libraries may wish to focus on Sections 2.2–2.3.1 and 5.4, plus all of Chapters 3, 7, and 8.

- Compiler writers will need to understand all of Chapters 2 and 3, plus Sections 4.5.2, 5.1, 5.3.1, 5.3.3, and 7.3–7.4.

Some readers may be surprised to find that the lecture contains no concrete performance results. This omission reflects a deliberate decision to focus on qualitative comparisons among algorithmic alternatives. Performance is obviously of great importance in the evaluation of synchronization mechanisms and concurrent data structures (and my papers are full of hard numbers), but the constants change with time, and they depend in many cases on characteristics of the specific application, language, operating system, and hardware at hand. When relative performance is in doubt, system designers would be well advised to benchmark the alternatives in their own particular environment.

Michael L. Scott
April 2013

Acknowledgments

This lecture has benefited from the feedback of many generous colleagues. Sarita Adve, Hans Boehm, Dave Dice, Maurice Herlihy, Mark Hill, Victor Luchangco, Paul McKenney, Maged Michael, Nir Shavit, and Mike Swift all read through draft material, and made numerous helpful suggestions for improvements. I am particularly indebted to Hans for his coaching on memory consistency models and to Victor for his careful vetting of Chapter 3. (The mistakes that remain are of course my own!) My thanks as well to the students of Mark's CS 758 course in the fall of 2012, who provided additional feedback. Finally, my admiration and thanks both to Mark and to Mike Morgan for their skillful shepherding of the Synthesis series, and for convincing me to undertake the project.

Michael L. Scott
April 2013

CHAPTER 1

Introduction

In computer science, as in real life, concurrency makes it much more difficult to reason about events. In a linear sequence, if E_1 occurs before E_2, which occurs before E_3, and so on, we can reason about each event individually: E_i begins with the state of the world (or the program) after E_{i-1}, and produces some new state of the world for E_{i+1}. But if the sequence of events $\{E_i\}$ is concurrent with some other sequence $\{F_i\}$, all bets are off. The state of the world prior to E_i can now depend not only on E_{i-1} and its predecessors, but also on some prefix of $\{F_i\}$.

Consider a simple example in which two threads attempt—concurrently—to increment a shared global counter:

```
thread 1:                       thread 2:
    ctr++                           ctr++
```

On any modern computer, the increment operation (ctr++) will comprise at least three separate instruction steps: one to load ctr into a register, a second to increment the register, and a third to store the register back to memory. This gives us a pair of concurrent sequences:

```
thread 1:                       thread 2:
1:   r := ctr                   1:   r := ctr
2:   inc r                      2:   inc r
3:   ctr := r                   3:   ctr := r
```

Intuitively, if our counter is initially 0, we should like it to be 2 when both threads have completed. If each thread executes line 1 before the other executes line 3, however, then both will store a 1, and one of the increments will be "lost."

The problem here is that concurrent sequences of events can *interleave* in arbitrary ways, many of which may lead to incorrect results. In this specific example, only two of the $\binom{6}{3} = 20$ possible interleavings—the ones in which one thread completes before the other starts—will produce the result we want.

Synchronization is the art of precluding interleavings that we consider incorrect. In a distributed (i.e., message-passing) system, synchronization is subsumed in communication: if thread T_2 receives a message from T_1, then in all possible execution interleavings, all the events performed by T_1 prior to its send will occur before any of the events performed by T_2 after its receive. In a shared-memory system, however, things are not so simple. Instead of exchanging messages, threads with shared memory communicate *implicitly* through loads and stores. Implicit communication gives the programmer substantially more flexibility in algorithm design, but it requires separate mechanisms for explicit synchronization. Those mechanisms are the subject of this lecture.

Significantly, the need for synchronization arises whenever operations are concurrent, regardless of whether they actually run in parallel. This observation dates from the earliest work in the field, led by Edsger Dijkstra [1965, 1968a, 1968b] and performed in the early 1960s. If a single processor core context switches among concurrent operations at arbitrary times, then while some interleavings of the underlying events may be less probable than they are with truly parallel execution, they are nonetheless *possible*, and a correct program must be synchronized to protect against any that would be incorrect. From the programmer's perspective, a multiprogrammed uniprocessor with preemptive scheduling is no easier to program than a multicore or multiprocessor machine.

A few languages and systems guarantee that only one thread will run at a time, and that context switches will occur only at well defined points in the code. The resulting execution model is sometimes referred to as "cooperative" multithreading. One might at first expect it to simplify synchronization, but the benefits tend not to be significant in practice. The problem is that potential context-switch points may be hidden inside library routines, or in the methods of black-box abstractions. Absent a programming model that attaches a true or false "may cause a context switch" tag to every method of every system interface, programmers must protect against unexpected interleavings by using synchronization techniques analogous to those of truly concurrent code.

Distribution

At the level of hardware devices, the distinction between shared memory and message passing disappears: we can think of a memory cell as a simple process that receives load and store messages from more complicated processes, and sends value and ok messages, respectively, in response. While theoreticians often think of things this way (the annual *PODC* [*Symposium on Principles of Distributed Computing*] and *DISC* [*International Symposium on Distributed Computing*] conferences routinely publish shared-memory algorithms), systems programmers tend to regard shared memory and message passing as fundamentally distinct. This lecture covers only the shared-memory case.

Concurrency and Parallelism

Sadly, the adjectives "concurrent" and "parallel" are used in different ways by different authors. For some authors (including the current one), two operations are *concurrent* if both have started and neither has completed; two operations are *parallel* if they may actually execute at the same time. Parallelism is thus an *implementation of concurrency*. For other authors, two operations are concurrent if there is no correct way to assign them an order in advance; they are parallel if their executions are independent of one another, so that any order is acceptable. An interactive program and its event handlers, for example, are concurrent with one another, but not parallel. For yet other authors, two operations that may run at the same time are considered concurrent (also called *task parallel*) if they execute different code; they are parallel if they execute the *same* code using different data (also called *data parallel*).

As it turns out, almost all synchronization patterns in real-world programs (i.e., all conceptually appealing constraints on acceptable execution interleaving) can be seen as instances of either *atomicity* or *condition synchronization*. Atomicity ensures that a specified sequence of instructions participates in any possible interleavings as a single, indivisible unit—that nothing else appears to occur in the middle of its execution. (Note that the very concept of interleaving is based on the assumption that underlying machine instructions are themselves atomic.) Condition synchronization ensures that a specified operation does not occur until some necessary precondition is true. Often, this precondition is the completion of some other operation in some other thread.

1.1   ATOMICITY

The example on p. 1 requires only atomicity: correct execution will be guaranteed (and incorrect interleavings avoided) if the instruction sequence corresponding to an increment operation executes as a single indivisible unit:

```
thread 1:                        thread 2:
    atomic                           atomic
        ctr++                            ctr++
```

The simplest (but not the only!) means of implementing atomicity is to force threads to execute their operations one at a time. This strategy is known as *mutual exclusion*. The code of an atomic operation that executes in mutual exclusion is called a *critical section*. Traditionally, mutual exclusion is obtained by performing acquire and release operations on an abstract data object called a *lock*:

```
lock L

thread 1:                        thread 2:
    L.acquire()                      L.acquire()
        ctr++                            ctr++
    L.release()                      L.release()
```

The acquire and release operations are assumed to have been implemented (at some lower level of abstraction) in such a way that (1) each is atomic and (2) acquire waits if the lock is currently held by some other thread.

In our simple increment example, mutual exclusion is arguably the only implementation strategy that will guarantee atomicity. In other cases, however, it may be overkill. Consider an operation that increments a specified element in an *array* of counters:

```
ctr_inc(i):
    L.acquire()
        ctr[i]++
    L.release()
```

If thread 1 calls ctr_inc(i) and thread 2 calls ctr_inc(j), we shall need mutual exclusion only if i = j. We can increase potential concurrency with a finer *granularity* of locking—for example, by declaring a separate lock for each counter, and acquiring only the one we need. In this example, the only downside is the space consumed by the extra locks. In other cases, fine-grain locking can introduce performance or correctness problems. Consider an operation designed to move n dollars from account i to account j in a banking program. If we want to use fine-grain locking (so unrelated transfers won't exclude one another in time), we need to acquire two locks:

```
move(n, i, j):
    L[i].acquire()
    L[j].acquire()            // (there's a bug here)
        acct[i] -= n
        acct[j] += n
    L[i].release()
    L[j].release()
```

If lock acquisition and release are expensive, we shall need to consider whether the benefit of concurrency in independent operations outweighs the cost of the extra lock. More significantly, we shall need to address the possibility of *deadlock*:

```
thread 1:                        thread 2:
    move(100, 2, 3)                  move(50, 3, 2)
```

If execution proceeds more or less in lockstep, thread 1 may acquire lock 2 and thread 2 may acquire lock 3 before either attempts to acquire the other. Both may then wait forever. The simplest solution in this case is to always acquire the lower-numbered lock first. In more general cases, if may be difficult to devise a static ordering. Alternative atomicity mechanisms—in particular, *transactional memory*, which we will consider in Chapter 9—attempt to achieve the concurrency of fine-grain locking without its conceptual complexity.

From the programmer's perspective, fine-grain locking is a means of implementing atomicity for large, complex operations using smaller (possibly overlapping) critical sections. The burden of ensuring that the implementation is correct (that it does, indeed, achieve deadlock-free atomicity for the large operations) is entirely the programmer's responsibility. The appeal of transactional memory is that it raises the level of abstraction, allowing the programmer to delegate this responsibility to some underlying system.

Whether atomicity is achieved through coarse-grain locking, programmer-managed fine-grain locking, or some form of transactional memory, the intent is that atomic regions appear to be indivisible. Put another way, any realizable execution of the program—any possible interleaving of its machine instructions—must be indistinguishable from (have the same externally visible behavior as) some execution in which the instructions of each atomic operation are contiguous in time, with no other instructions interleaved among them. As we shall see in Chapter 3, there are several possible ways to formalize this requirement, most notably *linearizability* and several variants on *serializability*.

1.2   CONDITION SYNCHRONIZATION

In some cases, atomicity is not enough for correctness. Consider, for example, a program containing a *work queue*, into which "producer" threads place tasks they wish to have performed, and from which "consumer" threads remove tasks they plan to perform. To preserve the structural integrity of the queue, we shall need each insert or remove operation to execute atomically. More than this, however, we shall need to ensure that a remove operation executes only when the queue is nonempty and (if the size of the queue is bounded) an insert operation executes only when the queue is nonfull:

```
Q.insert(d):                          Q.remove():
    atomic                                atomic
        await ¬Q.full()                       await ¬Q.empty()
        // put d in next empty slot            // return data from next full slot
```

In the synchronization literature, a concurrent queue (of whatever sort of objects) is sometimes called a *bounded buffer*; it is the canonical example of mixed atomicity and condition synchronization. As suggested by our use of the await *condition* notation above (notation we have not yet explained how to implement), the conditions in a bounded buffer can be specified at the beginning of the critical section. In other, more complex operations, a thread may need to perform nontrivial work within an atomic operation before it knows what condition(s) it needs to wait for. Since another thread will typically need to access (and modify!) some of the same data in order to make the condition true, a mid-operation wait needs to be able to "break" the atomicity of the surrounding operation in some well-defined way. In Chapter 7 we shall see that some synchronization mechanisms support only the simpler case of waiting at the beginning of a critical section; others allow conditions to appear anywhere inside.

In many programs, condition synchronization is also useful *outside* atomic operations—typically as a means of separating "phases" of computation. In the simplest case, suppose that a task to be performed in thread *B* cannot safely begin until some other task (data structure initialization, perhaps) has completed in thread *A*. Here *B* may spin on a Boolean *flag* variable that is initially false and that is set by *A* to true. In more complex cases, it is common for a program to go through a *series* of phases, each of which is internally parallel, but must complete in its entirety before the next phase can begin. Many simulations, for example, have this structure. For such programs, a *synchronization barrier*, executed by all threads at the end of every phase, ensures that all have arrived before any is allowed to depart.

It is tempting to suppose that atomicity (or mutual exclusion, at least) would be simpler to implement—or to model formally—than condition synchronization. After all, it could be thought of as a subcase: "wait until no other thread is currently in its critical section." The problem with this thinking is the scope of the condition. By standard convention, we allow conditions to consider only the values of variables, not the states of other threads. Seen in this light, atomicity is the more demanding concept: it requires agreement among *all* threads that their operations will avoid interfering with each other. And indeed, as we shall see in Section 3.3, atomicity is more difficult to implement, in a formal, theoretical sense.

1.3   SPINNING VS. BLOCKING

Just as synchronization patterns tend to fall into two main camps (atomicity and condition synchronization), so too do their implementations: they all employ *spinning* or *blocking*. Spinning is the simpler case. For condition synchronization, it takes the form of a trivial loop:

```
while ¬condition
    // do nothing (spin)
```

For mutual exclusion, the simplest implementation employs a special hardware instruction known as test_and_set (TAS). The TAS instruction, available on almost every modern machine, sets a specified Boolean variable to true and returns the previous value. Using TAS, we can implement a trivial *spin lock*:

```
type lock = bool := false

L.acquire():                        L.release():
    while ¬TAS(&L)                       L := false
        // spin
```

Here we have equated the acquisition of L with the act of *changing* it from false to true. The acquire operation repeatedly applies TAS to the lock until it finds that the previous value was false. As we shall see in Chapter 4, the trivial test_and_set lock has several major performance problems. It is, however, correct.

The obvious objection to spinning (also known as *busy-waiting*) is that it wastes processor cycles. In a multiprogrammed system it is often preferable to *block*—to yield the processor core to some other, runnable thread. The prior thread may then be run again later—either after some suitable interval of time (at which point it will check its condition, and possibly yield, again), or at some particular time when another thread has determined that the condition is finally true.

Processes, Threads, and Tasks

Like "concurrent" and "parallel," the terms "process," "thread," and "task" are used in different ways by different authors. In the most common usage (adopted here), a *thread* is an active computation that has the potential to share variables with other, concurrent threads. A *process* is a set of threads, together with the address space and other resources (e.g., open files) that they share. A *task* is a well-defined (typically small) unit of work to be accomplished—most often the closure of a subroutine with its parameters and referencing environment. Tasks are passive entities that may be executed by threads. They are invariably implemented at user level. The reader should beware, however, that this terminology is not universal. Many papers (particularly in theory) use "process" where we use "thread." Ada uses "task" where we use "thread." Mach uses "task" where we use "process." And some systems introduce additional words—e.g., "activation," "fiber," "filament," or "hart."

The software responsible for choosing which thread to execute when is known as a *scheduler*. In many systems, scheduling occurs at two different levels. Within the operating system, a kernel-level scheduler implements (kernel-level) threads on top of some smaller number of processor cores; within the user-level run-time system, a user-level scheduler implements (user-level) threads on top of some smaller number of kernel threads. At both levels, the code that implements threads (and synchronization) may present a library-style interface, composed entirely of subroutine calls; alternatively, the language in which the kernel or application is written may provide special syntax for thread management and synchronization, implemented by the compiler.

Certain issues are unique to schedulers at different levels. The kernel-level scheduler, in particular, is responsible for protecting applications from one another, typically by running the threads of each in a different address space; the user-level scheduler, for its part, may need to address such issues as non-conventional stack layout. To a large extent, however, the kernel and runtime schedulers have similar internal structure, and both spinning and blocking may be useful at either level.

While blocking saves cycles that would otherwise be wasted on fruitless re-checks of a condition or lock, it *spends* cycles on the context switching overhead required to change the running thread. If the average time that a thread expects to wait is less than twice the context-switch time, spinning will actually be faster than blocking. It is also the obvious choice if there is only one thread per core, as is sometimes the case in embedded or high-performance systems. Finally, as we shall see in Chapter 7, blocking (otherwise known as *scheduler-based synchronization*) must be built *on top of* spinning, because the data structures used by the scheduler itself require synchronization.

1.4   SAFETY AND LIVENESS

Whether based on spinning or blocking, a correct implementation of synchronization requires both *safety* and *liveness*. Informally, safety means that bad things never happen: we never have two threads in a critical section for the same lock at the same time; we never have all of the threads in the system blocked. Liveness means that good things eventually happen: if lock L is

---

Multiple Meanings of "Blocking"

"Blocking" is another word with more than one meaning. In this chapter, we are using it in an implementation-oriented sense, as a synonym for "de-scheduling" (giving the underlying kernel thread or hardware core to another user or kernel thread). In a similar vein, it is sometimes used in a "systems" context to refer to an operation (e.g., a "blocking" I/O request) that waits for a response from some other system component. In Chapter 3, we will use it in a more formal sense, as a synonym for "unable to make forward progress on its own." To a theoretician, a thread that is spinning on a condition that must be made true by some other thread is just as "blocked" as one that has given up its kernel thread or hardware core, and will not run again until some other thread tells the scheduler to resume it. Which definition we have in mind should usually be clear from context.

free and at least one thread is waiting for it, some thread eventually acquires it; if queue Q is nonempty and at least one thread is waiting to remove an element, some thread eventually does.

A bit more formally, for a given program and input, running on a given system, safety properties can always be expressed as predicates P on reachable system states S—that is, $\forall S[P(S)]$. Liveness properties require at least one extra level of quantification: $\forall S[P(S) \rightarrow \exists T[Q(T)]]$, where T is a subsequent state in the *same execution* as S, and Q is some other predicate on states. From a practical perspective, liveness properties tend to be harder than safety to ensure—or even to define; from a formal perspective, they tend to be harder to prove.

Livelock freedom is one of the simplest liveness properties. It insists that threads not execute forever without making forward progress. In the context of locking, this means that if L is free and thread T has called L.acquire(), there must exist some bound on the number of instructions T can execute before *some* thread acquires L. *Starvation freedom* is stronger. Again in the context of locks, it insists that if every thread that acquires L eventually releases it, and if T has called L.acquire(), there must exist some bound on the number of instructions T can execute before acquiring L itself. Still stronger notions of *fairness* among threads can also be defined; we consider these briefly in Section 3.2.2.

Most of our discussions of correctness will focus on safety properties. Interestingly, *deadlock freedom*, which one might initially imagine to be a matter of liveness, is actually one of safety: because deadlock can be described as a predicate that takes the current system state as input, deadlock freedom simply insists that the predicate be false in all reachable states.

CHAPTER 2

Architectural Background

The correctness and performance of synchronization algorithms depend crucially on architectural details of multicore and multiprocessor machines. This chapter provides an overview of these details. It can be skimmed by those already familiar with the subject, but should probably not be skipped in its entirety: the implications of store buffers and directory-based coherence on synchronization algorithms, for example, may not be immediately obvious, and the semantics of synchronizing instructions (ordered accesses, memory fences, and *read-modify-write* instructions) may not be universally familiar.

The chapter is divided into three main sections. In the first, we consider the implications for parallel programs of caching and coherence protocols. In the second, we consider *consistency*—the degree to which accesses to different memory locations can or cannot be assumed to occur in any particular order. In the third, we survey the various read-modify-write instructions—test_and_set and its cousins—that underlie most implementations of atomicity.

2.1 CORES AND CACHES: BASIC SHARED-MEMORY ARCHITECTURE

Figures 2.1 and 2.2 depict two of the many possible configurations of processors, cores, caches, and memories in a modern parallel machine. In a so-called *symmetric* machine, all memory banks are equally distant from every processor core. Symmetric machines are sometimes said to have a *uniform memory access* (UMA) architecture. More common today are *nonuniform memory access* (NUMA) machines, in which each memory bank is associated with a processor (or in some cases with a multi-processor *node*), and can be accessed by cores of the local processor more quickly than by cores of other processors.

As feature sizes continue to shrink, the number of cores per processor can be expected to increase. As of this writing, the typical desk-side machine has 1–4 processors with 2–16 cores each. Server-class machines are architecturally similar, but with the potential for many more processors. Small machines often employ a symmetric architecture for the sake of simplicity. The physical distances in larger machines often motivate a switch to NUMA architecture, so that local memory accesses, at least, can be relatively fast.

On some machines, each core may be multithreaded—capable of executing instructions from more than one thread at a time (current per-core thread counts range from 1–8). Each core typically has a private level-1 (L1) cache, and shares a level-2 cache with other cores in its local *cluster*. Clusters on the same processor of a symmetric machine then share a common L3

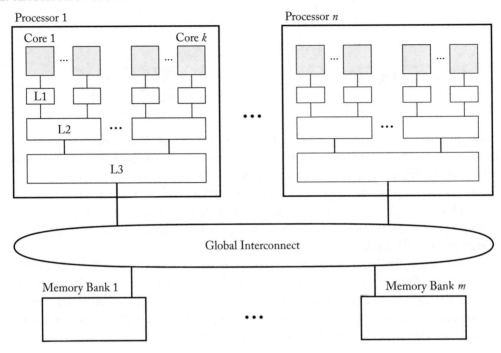

Figure 2.1: Typical symmetric (uniform memory access—UMA) machine. Numbers of components of various kinds, and degree of sharing at various levels, differs across manufacturers and models.

cache. Each cache holds a temporary copy of data currently in active use by cores above it in the hierarchy, allowing those data to be accessed more quickly than they could be if kept in memory. On a NUMA machine in which the L2 connects directly to the global interconnect, the L3 may sometimes be thought of as "belonging" to the memory.

In a machine with more than one processor, the global interconnect may have various topologies. On small machines, broadcast buses and crossbars are common; on large machines, a network of point-to-point links is more common. For synchronization purposes, broadcast has the side effect of imposing a total order on all inter-processor messages; we shall see in Section 2.2 that this simplifies the design of concurrent algorithms—synchronization algorithms in particular. Ordering is sufficiently helpful, in fact, that some large machines (notably those sold by Oracle) employ two different global networks: one for data requests, which are small, and benefit from ordering, and the other for replies, which require significantly more aggregate bandwidth, but do not need to be ordered.

As the number of cores per processor increases, on-chip interconnects—the connections among the L2 and L3 caches in particular—can be expected to take on the complexity of current global interconnects. Other forms of increased complexity are also likely, including, perhaps,

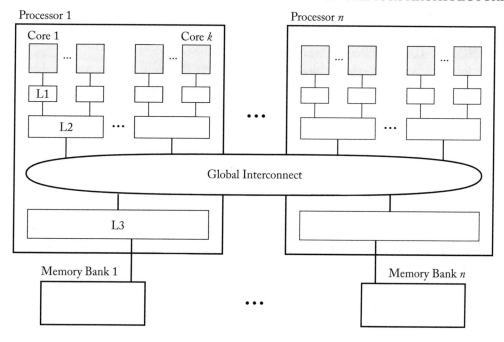

Figure 2.2: Typical nonuniform memory access (NUMA) machine. Again, numbers of components of various kinds, and degree of sharing at various levels, differs across manufacturers and models.

additional levels of caching, non-hierarchical topologies, and heterogeneous implementations or even instruction sets among cores.

The diversity of current and potential future architectures notwithstanding, multilevel caching has several important consequences for programs on almost any modern machine; we explore these in the following subsections.

2.1.1 TEMPORAL AND SPATIAL LOCALITY

In both sequential and parallel programs, performance can usually be expected to correlate with the temporal and spatial locality of memory references. If a given location l is accessed more than once by the same thread (or perhaps by different threads on the same core or cluster), performance is likely to be better if the two references are close together in time (temporal locality). The benefit stems from the fact that l is likely still to be in cache, and the second reference will be a hit instead of a miss. Similarly, if a thread accesses location l_2 shortly after l_1, performance is likely to be better if the two locations have nearby addresses (spatial locality). Here the benefit stems from the fact that l_1 and l_2 are likely to lie in the same *cache line*, so l_2 will have been loaded into cache as a side effect of loading l_1.

On current machines, cache line sizes typically vary between 32 and 512 bytes. There has been a gradual trend toward larger sizes over time. Different levels of the cache hierarchy may also use different sizes, with lines at lower levels typically being larger.

To improve temporal locality, the programmer must generally restructure algorithms, to change the order of computations. Spatial locality is often easier to improve—for example, by changing the layout of data in memory to co-locate items that are frequently accessed together, or by changing the order of traversal in multidimensional arrays. These sorts of optimizations have long been an active topic of research, even for sequential programs—see, for example, the texts of Muchnick [1997, Chap. 20] or Allen and Kennedy [2002, Chap. 9].

2.1.2   CACHE COHERENCE

On a single-core machine, there is a single cache at each level, and while a given block of memory may be present in more than one level of the memory hierarchy, one can always be sure that the version closest to the top contains up-to-date values. (With a *write-through* cache, values in lower levels of the hierarchy will never be out of date by more than some bounded amount of time. With a *write-back* cache, values in lower levels may be arbitrarily stale—but harmless, because they are hidden by copies at higher levels.)

On a shared-memory parallel system, by contrast—unless we do something special—data in upper levels of the memory hierarchy may no longer be up-to-date if they have been modified by some thread on another core. Suppose, for example, that threads on Cores 1 and k in Figure 2.1 have both been reading variable x, and each has retained a copy in its L1 cache. If the thread on Core 1 then modifies x, even if it writes its value through (or back) to memory, how do we prevent the thread on Core k from continuing to read the stale copy?

A *cache-coherent* parallel system is one in which (1) changes to data, even when cached, are guaranteed to become visible to all threads, on all cores, within a bounded (and typically small) amount of time; and (2) changes to the same location are seen in the same order by all threads. On almost all modern machines, coherence is achieved by means of an *invalidation-based cache coherence protocol*. Such a protocol, operating across the system's cache controllers, maintains the invariant that there is at most one writable copy of any given cache line anywhere in the system—and, if the number is one and not zero, there are no read-only copies.

Algorithms to maintain cache coherence are a complex topic, and the subject of ongoing research (for an overview, see the lecture of Sorin et al. [2011], or the more extensive [if dated] coverage of Culler and Singh [1998, Chaps. 5, 6, & 8]). Most protocols in use today descend from the four-state protocol of Goodman [1983]. In this protocol (using modern names for the states), each line in each cache is either *invalid* (meaning it currently holds no data block), *shared* (read-only), *exclusive* (written exactly once, and up-to-date in memory), or *modified* (written more than once, and in need of write-back).

To maintain the at-most-one-writable-copy invariant, the coherence protocol arranges, on any write to an *invalid* or *shared* line, to invalidate (evict) any copies of the block in all other caches in the system. In a system with a broadcast-based interconnect, invalidation is straightforward. In a system with point-to-point connections, the coherence protocol typically maintains some sort of *directory* information that allows it to find all other copies of a block.

2.1.3   PROCESSOR (CORE) LOCALITY

On a single-core machine, misses occur on an initial access (a "cold miss") and as a result of limited cache capacity or associativity (a "conflict miss").[1] On a cache-coherent machine, misses may also occur because of the need to maintain coherence. Specifically, a read or write may miss because a previously cached block has been written by some other core, and has reverted to *invalid* state; a write may also miss because a previously *exclusive* or *modified* block has been read by some other core, and has reverted to *shared* state.

Absent program restructuring, coherence misses are inevitable if threads on different cores access the same datum (and at least one of them writes it) at roughly the same point in time. In addition to temporal and spatial locality, it is therefore important for parallel programs to exhibit good *thread locality*: as much possible, a given datum should be accessed by only one thread at a time.

Coherence misses may sometimes occur when threads are accessing *different* data, if those data happen to lie in the same cache block. This *false sharing* can often be eliminated—yielding a major speed improvement—if data structures are *padded* and *aligned* to occupy an integral number of cache lines. For busy-wait synchronization algorithms, it is particularly important to minimize the extent to which different threads may spin on the same location—or locations in the same cache block. Spinning with a write on a shared location—as we did in the test_and_set lock of Section 1.3, is particularly deadly: each such write leads to interconnect traffic proportional to the number of other spinning threads. We will consider these issues further in Chapter 4.

---

[1]A cache is said to be *k-way associative* if its indexing structure permits a given block to be cached in any of *k* distinct locations. If *k* is 1, the cache is said to be *direct mapped*. If a block may be held in any line, the cache is said to be *fully associative*.

No-remote-caching Multiprocessors

Most of this lecture assumes a shared-memory multiprocessor with global (distributed) cache coherence, which we have contrasted with machines in which message passing provides the only means of interprocessor communication. There is an intermediate option. Some NUMA machines (notably many of the offerings from Cray, Inc.) support a single global address space, in which any processor can access any memory location, but remote locations cannot be cached. We may refer to such a machine as a no-remote-caching (NRC-NUMA) multiprocessor. (Globally cache coherent NUMA machines are sometimes known as CC-NUMA.) Any access to a location in some other processor's memory will traverse the interconnect of an NRC-NUMA machine. Assuming the hardware implements cache coherence *within* each node—in particular, between the local processor(s) and the network interface—memory will still be globally coherent. For the sake of performance, however, system and application programmers will need to employ algorithms that minimize the number of remote references.

2.2   MEMORY CONSISTENCY

On a single-core machine, it is relatively straightforward to ensure that instructions appear to complete in execution order. Ideally, one might hope that a similar guarantee would apply to parallel machines—that memory accesses, system-wide, would appear to constitute an interleaving (in execution order) of the accesses of the various cores. For several reasons, this sort of *sequential consistency* [Lamport, 1979] imposes nontrivial constraints on performance. Most real machines implement a more *relaxed* (i.e., potentially inconsistent) memory model, in which accesses by different threads, or to different locations by the same thread, may appear to occur "out of order" from the perspective of threads on other cores. When consistency is required, programmers (or compilers) must employ special *synchronizing instructions* that are more strongly ordered than other, "ordinary" instructions, forcing the local core to wait for various classes of potentially in-flight events. Synchronizing instructions are an essential part of synchronization algorithms on any non-sequentially consistent machine.

2.2.1   SOURCES OF INCONSISTENCY

Inconsistency is a natural result of common architectural features. In an *out-of-order* processor, for example—one that can execute instructions in any order consistent with (thread-local) data dependences—a write must be held in the *reorder buffer* until all instructions that precede it in program order have completed. Likewise, since almost any modern processor can generate a burst of store instructions faster than the underlying memory system can absorb them, even writes that are logically ready to commit may need to be buffered for many cycles. The structure that holds these writes is known as a *store buffer*.

When executing a load instruction, a core checks the contents of its reorder and store buffers before forwarding a request to the memory system. This check ensures that the core always sees its own recent writes, even if they have not yet made their way to cache or memory. At the same time, a load that accesses a location that has *not* been written recently may make its way to memory before logically previous instructions that wrote to other locations. This fact is harmless on a uniprocessor, but consider the implications on a parallel machine, as shown in Figure 2.3. If the write to x is delayed in thread 1's store buffer, and the write to y is similarly delayed in thread 2's store buffer, then both threads may read a zero at line 2, suggesting that line 2 of thread 1 executes before line 1 of thread 2, and line 2 of thread 2 executes before line 1 of thread 1. When combined with program order (line 1 in each thread should execute before line 2 in the same thread), this gives us an apparent "ordering loop," which "should" be logically impossible.

Similar problems can occur deeper in the memory hierarchy. A modern machine can require several hundred cycles to service a miss that goes all the way to memory. At each step along the way (core to L1, …, L3 to bus, …) pending requests may be buffered in a queue. If multiple requests may be active simultaneously (as is common, at least, on the global interconnect), and if some requests may complete more quickly than others, then memory accesses may appear to be reordered. So long as accesses to the *same* location (by the same thread) are forced to occur in order,

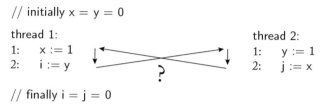

```
// initially x = y = 0

thread 1:                          thread 2:
1:    x := 1                        1:    y := 1
2:    i := y                        2:    j := x

// finally i = j = 0
```

Figure 2.3: An apparent ordering loop.

single-threaded code will run correctly. On a multiprocessor, however, sequential consistency may again be violated.

On a NUMA machine, or a machine with a topologically complex interconnect, differing distances among locations provide additional sources of circular ordering. If variable x in Figure 2.3 is close to thread 2 but far from thread 1, and y is close to thread 1 but far from thread 2, the reads on line 2 can easily complete before the writes on line 1, even if all accesses are inserted into the memory system in program order. With a topologically complex interconnect, the cache coherence protocol itself may introduce variable delays—e.g., to dispatch invalidation requests to the various locations that may need to change the state of a local cache line, and to collect acknowledgments. Again, these differing delays may allow line 2 of the example—in both threads—to complete before line 1.

In all the explanations of Figure 2.3, the ordering loop results from reads *bypassing* writes—executing in-order (write-then-read) from the perspective of the issuing core, but out of order (read-then-write) from the perspective of the memory system—or of threads on other cores. On NUMA or topologically complex machines, it may also be possible for reads to bypass reads, writes to bypass reads, or writes to bypass writes. Worse, circularity may arise even without bypassing—i.e., even when every thread executes its own instructions in strict program order. Consider the "independent reads of independent writes" (IRIW) example shown in Figure 2.4. If thread 1 is close to thread 2 but far from thread 3, and thread 4 is close to thread 3 but far from thread 2, the reads on line 1 in threads 2 and 3 may see the new values of x and y, while the reads on line 2 see the old. Here the problem is not bypassing, but a lack of *write atomicity*—one thread sees the value written by a store and another thread *subsequently* sees the value prior to the store. Many other examples of unintuitive behavior permitted by modern hardware can be found in the literature [Adve and Gharachorloo, 1996, Adve et al., 1999, Boehm and Adve, 2008, Manson et al., 2005].

2.2.2   SPECIAL INSTRUCTIONS TO ORDER MEMORY ACCESS

If left unaddressed, memory inconsistency can easily derail attempts at synchronization. Consider the flag-based programming idiom illustrated in Figure 2.5. If foo can never return zero, a programmer might naively expect that thread 2 will never see a divide-by-zero error at line 3. If

// initially x = y = 0

thread 1:   thread 2:   thread 3:   thread 4:
1:    x := 1 ──────→ 1:   x2 := x   1:   y3 := y ←────── 1:   y := 1
                   2:    y2 := y   ?   2:   x3 := x

// finally y2 = x3 = 0 and x2 = y3 = 1

Figure 2.4: Independent reads of independent writes (IRIW). If the writes of threads 1 and 4 propagate to different places at different speeds, we can see a ordering loop even if instructions from the same thread never bypass one another.

// initially x = f = 0

thread 1:   thread 2:
1:    x := foo()   1:   while f = 0
2:    f := 1   2:   // spin
3:                                  3:   y := 1/x

Figure 2.5: A simple example of flag-based synchronization. To avoid a spurious error, the update to x must be visible to thread 2 before the update to f.

the write at line 2 in thread 1 can bypass the write in line 1, however, thread 2 may read x too early, and see a value of zero. Similarly, if the read of x at line 3 in thread 2 can bypass the read of f in line 1, a divide-by-zero may again occur, even if the writes in thread 1 complete in order. (While thread 2's read of x is separated from the read of f by a conditional test, the second read may still issue before the first completes, if the branch predictor guesses that the loop will never iterate.)

Compilers Also Reorder Instructions

While this chapter focuses on architectural issues, it should be noted that compilers also routinely reorder instructions. In any program not written in machine code, compilers perform a variety of optimizations in an attempt to improve performance. Simple examples include reordering computations to expose and eliminate redundancies, hoisting invariants out of loops, and "scheduling" instructions to minimize processor pipeline bubbles. Such optimizations are legal so long as they respect control and data dependences within a single thread. Like the hardware optimizations discussed in this section, compiler optimizations can lead to inconsistent behavior when more than one thread is involved. As we shall see in Section 3.4, languages designed for concurrent programming must provide a *memory model* that explains allowable behavior, and some set of primitives—typically special synchronization operations or reads and writes of special atomic variables—that serve to order accesses at the language level.

Any machine that is not sequentially consistent will provide special instructions that allow the programmer to force consistent ordering in situations in which it matters, but in which the hardware might not otherwise guarantee it. Perhaps the simplest such instruction is a synchronizing access (typically a special load or store) that is guaranteed to be both locally and globally ordered. Here "locally ordered" means that the synchronizing access will appear to occur after any preceding ordinary accesses in its own thread, and before any subsequent ordinary accesses in its thread, from the perspective of all threads. "Globally ordered" means that the synchronizing access will appear to occur in some consistent, total order with respect to all other synchronizing instructions in the program, from the perspective of all threads. (As part of global ordering, we also require that synchronizing accesses by the same thread appear to occur in program order, from the perspective of all threads.) At the hardware level, global ordering is typically achieved by means of write atomicity: the cache coherence protocol prevents a load from returning the value written by a synchronizing store until it verifies that no load elsewhere in the machine can ever again return the previous value.

To avoid the spurious error in Figure 2.5, it is sufficient (though not necessary) to use fully ordered accesses to f in both threads, thereby ensuring that thread 1's update of x happens before its update of f, and thread 2's read of x happens after it sees the change to f. Alternatively, ordering could be ensured by inserting a *full fence* instruction between lines 1 and 2 in thread 1, and between lines 2 and 3 in thread 2. A full fence doesn't read or write memory itself, but it ensures that all preceding memory accesses in its thread appear to occur before all subsequent memory accesses in its thread, from the perspective of other threads.

In Figure 2.3, circularity could be avoided by using fully ordered stores for the line-1 writes (in both threads) or by using fully ordered loads for the line-2 reads (in both threads). Alternatively, we could insert full fences between lines 1 and 2 in both threads. In Figure 2.4, however, global ordering (i.e., write atomicity) is really all that matters; to ensure it, we must use synchronizing stores in threads 1 and 4. Synchronizing loads in threads 2 and 3—or fences between the loads—will not address the problem: absent write atomicity, it is possible for thread 1's write to appear to happen before the entire set of reads—and thread 4's write after—from the perspective of thread 2, and vice versa for thread 3.

Barriers Everywhere

Fences are sometimes known as *memory barriers*. Sadly, the word *barrier* is heavily overloaded. As noted in Section 1.2 (and explored in more detail in Section 5.2), it is the name of a synchronization mechanism used to separate program phases. In the programming language community, it refers to code that must be executed when changing a pointer, in order to maintain bookkeeping information for the garbage collector. In a similar vein, it sometimes refers to code that must be executed when reading or writing a shared variable inside an atomic *transaction*, in order to detect and recover from speculation failures (we discuss this code in Chapter 9, but without referring to it as a "barrier"). The intended meaning is usually clear from context, but may be confusing to readers who are familiar with only some of the definitions.

On many machines, fully ordered synchronizing instructions turn out to be quite expensive—tens or even hundreds of cycles. Moreover, in many cases—including those described above—full ordering is more than we need for correct behavior. Architects therefore often provide a variety of weaker synchronizing instructions. These may or may not be globally ordered, and may prevent some, but not all, local bypassing. As we shall see in Section 2.2.3, the details vary greatly from one machine architecture to another. Moreover, behavior is often defined not in terms of the orderings an instruction guarantees among memory accesses, but in terms of the *re*orderings it inhibits in the processor core, the cache subsystem, or the interconnect.

Unfortunately, there is no obvious, succinct way to specify minimal ordering requirements in parallel programs. Neither synchronizing accesses nor fences, for example, allow us to order two individual accesses with respect to one another (and not with respect to anything else), if that is all that is really required. In an attempt to balance simplicity and clarity, the examples in this lecture use a notation inspired by (but simpler than) the `atomic` operations of C++'11. Using this notation, we will sometimes over-constrain our algorithms, but not egregiously.

A summary of our notation, and of the memory model behind it, can be found in Table 2.1. To specify local ordering, each synchronizing instruction admits an optional annotation of the form $P\|S$, indicating that the instruction is ordered with respect to preceding (P) and/or subsequent (S) read and write accesses in its thread ($P, S \subset \{R, W\}$). So, for example, f.store(1, W$\|$) might be used in Figure 2.5 at line 2 of thread 1 to order the (synchronizing) store to f after the (ordinary) write to x, and f.load($\|$RW) might be used at line 1 of thread 2 to order the (synchronizing) load of f before both the (ordinary) read of x and any other subsequent reads and writes. Similarly, fence(RW$\|$RW) would indicate a full fence, ordered globally with respect to all other synchronizing instructions and locally with respect to all preceding and subsequent ordinary accesses in its thread.

We will assume that synchronizing instructions inhibit reordering not only by the hardware (processor, cache, or interconnect), but also by the compiler or interpreter. Compiler writers or assembly language programmers interested in porting our pseudocode to some concrete machine will need to restrict their code improvement algorithms accordingly, and issue appropriate synchronizing instructions for the hardware at hand. Beginning guidance can be found in Doug Lea's on-line "Cookbook for Compiler Writers" [2001].

To determine the need for synchronizing instructions in the code of a given synchronization algorithm, we shall need to consider both the correctness of the algorithm itself and the semantics it is intended to provide to the rest of the program. The acquire operation of Peterson's two-thread spin lock [1981], for example, employs synchronizing stores to arbitrate between competing threads, but this ordering is not enough to prevent a thread from reading or writing shared data before the lock has actually been acquired—or after it has been released. For that, one needs accesses or fences with local $\|$RW and RW$\|$ ordering (code in Section 4.1).

Fortunately for most programmers, memory ordering details are generally of concern only to the authors of synchronization algorithms and low-level concurrent data structures, which

Table 2.1: Understanding the pseudocode

Throughout the remainder of this book, pseudocode will be set in sans serif font code (code in real programming languages will be set in typewriter font). We will use the term *synchronizing instruction* to refer to explicit loads and stores, fences, and atomic read-modify-write (fetch_and_Φ) operations (listed in Table 2.2). Other memory accesses will be referred to as "ordinary." We will assume the following:

coherence

> All accesses (ordinary and synchronizing) to any given location appear to occur in some single, total order from the perspective of all threads.

global order

> There is a global, total order on synchronizing instructions (to all locations, by all threads). Within this order, instructions issued by the same thread occur in program order.

program order

> Ordinary accesses appear to occur in program order from the perspective of the issuing thread, but may bypass one another in arbitrary ways from the perspective of other threads.

local order

> Ordinary accesses may also bypass synchronizing instructions, except when forbidden by an ordering annotation ({R,W}‖{R,W}) on the synchronizing instruction.

values read

> A read instruction will return the value written by the most recent write (to the same location) that is ordered before the read. It may also, in some cases, return the value written by an unordered write. More detail on memory models can be found in Section 3.4.

may need to be re-written (or at least re-tuned) for each target architecture. Programmers who use these algorithms correctly are then typically assured that their programs will behave as if the hardware were sequentially consistent (more on this in Section 3.4), and will port correctly across machines.

Identifying a minimal set of ordering instructions to ensure the correctness of a given algorithm on a given machine is a difficult and error-prone task. It has traditionally been performed by hand, though there has been promising recent work aimed at verifying the correctness of a given set of fences [Burckhardt et al., 2007], or even at inferring them directly [Kuperstein et al., 2010].

2.2.3   EXAMPLE ARCHITECTURES

A few multiprocessors (notably, those built around the c. 1996 MIPS R10000 processor [Yeager, 1996]) have been defined to be sequentially consistent. A few others (notably, the HP PA-RISC) have been implemented with sequential consistency, even though the documentation permitted something more relaxed. Hill [1998] has argued that the overhead of sequential consistency need not be onerous, particularly in comparison to its conceptual benefits. Most machines today, however, fall into two broad classes of more relaxed alternatives. On the SPARC, x86 (both 32- and

64-bit), and IBM z Series, reads are allowed to bypass writes, but R‖R, R‖W, and W‖W orderings are all guaranteed to be respected by the hardware, and writes are always globally ordered. Special instructions—synchronizing accesses or fences—are required only when the programmer must ensure that a write and a subsequent read complete in program order. On ARM, POWER, and IA-64 (Itanium) machines, all four combinations of local bypassing are possible: special instructions must be used whenever ordering is required. Moreover, on ARM and POWER, ordinary writes are not guaranteed to be atomic.

We will refer to hardware-level memory models in the SPARC/x86/z camp using the SPARC term *TSO* (Total Store Order). We will refer to the other machines as "more relaxed." On TSO machines, R‖R, R‖W, and W‖W annotations can all be elided from our code. On more relaxed machines, they must be implemented with appropriate machine instructions. It should be emphasized that there are significant differences among machines within a given camp—in the default ordering, the available synchronizing instructions, and the details of corner cases. A full explanation is well beyond what we can cover here. For a taste of the complexities involved, see Sewell et al.'s attempt [2010] to formalize the behavior specified informally in Intel's architecture manual [2011, Vol. 3, Sec. 8.2].

A few machines—notably the Itanium and ARM v8 (but not v7)—provide explicit synchronizing access (load and store) instructions. Most machines provide separate fence instructions, which do not, themselves, modify memory. Though less common, synchronizing accesses have two principal advantages. First, they can achieve write atomicity, thereby precluding the sort of circularity shown in Figure 2.4. Second, they allow an individual access (the load or store itself) to be ordered with respect to preceding or subsequent ordinary accesses. A fence, by contrast, must order preceding or subsequent ordinary accesses with respect to *all* reads or writes in the opposite direction.

One particular form of local ordering is worthy of special mention. As noted in Section 2.2.2, a lock acquire operation must ensure that a thread cannot read or write shared data until it has actually acquired the lock. The appropriate guarantee will typically be provided by a ‖RW load or, more conservatively, a subsequent R‖RW fence. In a similar vein, a lock release must ensure that all reads and writes within the critical section have completed before the lock is actually released. The appropriate guarantee will typically be provided by a RW‖ store or, more conservatively, a preceding RW‖W fence. These combinations are common enough that they are sometimes referred to as *acquire* and *release* orderings. They are used not only for mutual exclusion, but for most forms of condition synchronization as well. The IA-64 (Itanium), ARM v8, and several research machines—notably the Stanford Dash [Lenoski et al., 1992]—support acquire and release orderings directly in hardware. In the mutual exclusion case, particularly when using synchronizing accesses rather than fences, acquire and release orderings allow work to "migrate" into a critical section both from above (prior to the lock acquire) and from below (after the lock release). They do *not* allow work to migrate *out* of a critical section in either direction.

Table 2.2:   Common atomic (read-modify-write) instructions

test_and_set
> bool TAS(bool *a): atomic { t := *a; *a := true; return t }

swap
> word Swap(word *a, word w): atomic { t := *a; *a := w; return t }

fetch_and_increment
> int FAI(int *a): atomic { t := *a; *a := t + 1; return t }

fetch_and_add
> int FAA(int *a, int n): atomic { t := *a; *a := t + n; return t }

compare_and_swap
> bool CAS(word *a, word old, word new):
>> atomic { t := (*a = old); if (t) *a := new; return t }

load_linked / store_conditional
> word LL(word *a): atomic { remember a; return *a }
> bool SC(word *a, word w):
>> atomic { t := (a is remembered, and has not been evicted since LL)
>>> if (t) *a := w; return t }

2.3   ATOMIC PRIMITIVES

To facilitate the construction of synchronization algorithms and concurrent data structures, most modern architectures provide instructions capable of updating (i.e., reading *and* writing) a memory location as a single atomic operation. We saw a simple example—the test_and_set instruction (TAS)—in Section 1.3. A longer list of common instructions appears in Table 2.2. Note that for each of these, when it appears in our pseudocode, we permit an optional, final argument that indicates local ordering constraints. CAS(a, old, new, W‖), for example, indicates a CAS instruction that is ordered after all preceding write accesses in its thread.

Originally introduced on mainframes of the 1960s, TAS and Swap are still available on several modern machines, among them the x86 and SPARC. FAA and FAI were introduced for "combining network" machines of the 1980s [Kruskal et al., 1988]. They are uncommon in hardware today, but frequently appear in algorithms in the literature. The semantics of TAS, Swap, FAI, and FAA should all be self-explanatory. Note that they all return the value of the target location *before* any change was made.

CAS was originally introduced in the 1973 version of the IBM 370 architecture [Brown and Smith, 1975, Gifford et al., 1987, IBM, 1975]. It is also found on modern x86, IA-64 (Itanium), and SPARC machines. LL/SC was originally proposed for the S-1 AAP Multiprocessor at Lawrence Livermore National Laboratory [Jensen et al., 1987]. It is also found on modern POWER, MIPS, and ARM machines. CAS and LL/SC are *universal* primitives, in a sense we will define formally in Section 3.3. In practical terms, we can use them to build efficient simulations of arbitrary (single-word) read-modify-write (fetch_and_Φ) operations (including all the other operations in Table 2.2).

CAS takes three arguments: a memory location, an old value that is expected to occupy that location, and a new value that should be placed in the location if indeed the old value is currently there. The instruction returns a Boolean value indicating whether the replacement occurred successfully. Given CAS, fetch_and_Φ can be written as follows, for any given function Φ:

```
1:   word fetch_and_Φ(function Φ, word *w):
2:       word old, new
3:       repeat
4:           old := *w
5:           new := Φ(old)
6:       until CAS(w, old, new)
7:       return old
```

In effect, this code computes $\Phi(*w)$ *speculatively*, and then updates w atomically if its value has not changed since the speculation began. The only way the CAS can fail to perform its update (and return false at line 6) is if some other thread has recently modified w. If several threads attempt to perform a fetch_and_Φ on w simultaneously, one of them is guaranteed to succeed, and the system as a whole will make forward progress. This guarantee implies that fetch_and_Φ operations implemented with CAS are *nonblocking* (more specifically, *lock free*), a property we will consider in more detail in Section 3.2.

One problem with CAS, from an architectural point of view, is that it combines a load and a store into a single instruction, which complicates the implementation of pipelined processors. LL/SC was designed to address this problem. In the fetch_and_Φ idiom above, it replaces the load at line 4 with a special instruction that has the side effect of "tagging" the associated cache line so that the processor will "notice" any subsequent eviction of the line. A subsequent SC will then succeed only if the line is still present in the cache:

```
word fetch_and_Φ(function Φ, word *w):
    word old, new
    repeat
        old := LL(w)
        new := Φ(old)
    until SC(w, new)
    return old
```

Here any argument for forward progress requires an understanding of why SC might fail. Details vary from machine to machine. In all cases, SC is guaranteed to fail if another thread has modified *w (the location pointed at by w) since the LL was performed. On most machines, SC will also fail if a hardware interrupt happens to arrive in the post-LL window. On some machines, it will fail if the cache suffers a capacity or conflict miss, or if the processor mispredicts a branch. To avoid deterministic, spurious failure, the programmer may need to limit (perhaps severely) the types of instructions executed between the LL and SC. If unsafe instructions are required in order to compute the function Φ, one may need a hybrid approach:

```
1:    word fetch_and_Φ(function Φ, word *w):
2:        word old, new
3:        repeat
4:            old := *w
5:            new := Φ(old)
6:        until LL(w) = old && SC(w, new)
7:        return old
```

In effect, this code uses LL and SC at line 6 to emulate CAS.

2.3.1   THE ABA PROBLEM

While both CAS and LL/SC appear in algorithms in the literature, the former is quite a bit more common—perhaps because its semantics are self-contained, and do not depend on the implementation-oriented side effect of cache-line tagging. That said, CAS has one significant disadvantage from the programmer's point of view—a disadvantage that LL/SC avoids.

Because it chooses whether to perform its update based on the value in the target location, CAS may succeed in situations where the value has changed (say from A to B) and then changed back again (from B to A) [IBM, 1975, 1983, Treiber, 1986]. In some algorithms, such a change and restoration is harmless: it is still acceptable for the CAS to succeed. In other algorithms, incorrect behavior may result. This possibility, often referred to as the *ABA problem*, is particularly worrisome in pointer-based algorithms. Consider the following (buggy!) code to manipulate a linked-list stack:

Emulating CAS

Note that while LL/SC can be used to emulate CAS, the emulation requires a loop to deal with spurious SC failures. This issue was recognized explicitly by the designers of the C++'11 atomic types and operations, who introduced two variants of CAS. The atomic_compare_exchange_strong operation has the semantics of hardware CAS: it fails only if the expected value was not found. On an LL/SC machine, it is implemented with a loop. The atomic_compare_exchange_weak operation admits the possibility of spurious failure: it has the interface of CAS, but is implemented *without* a loop on an LL/SC machine.

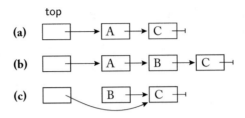

Figure 2.6: The ABA problem in a linked-list stack.

```
1: void push(node** top, node* new):        1: node* pop(node** top):
2:    node* old                              2:    node* old, new
3:    repeat                                 3:    repeat
4:        old := *top                        4:        old := *top
5:        new→next := old                    5:        if old = null return null
6:    until CAS(top, old, new)               6:        new := old→next
                                             7:    until CAS(top, old, new)
                                             8:    return old
```

Figure 2.6 shows one of many problem scenarios. In (a), our stack contains the elements A and C. Suppose that thread 1 begins to execute pop(&top), and has completed line 6, but has yet to reach line 7. If thread 2 now executes a (complete) pop(&top) operation, followed by push(&top, &B) and then push(&top, &A), it will leave the stack as shown in (b). If thread 1 now continues, its CAS will succeed, leaving the stack in the broken state shown in (c).

The problem here is that top changed between thread 1's load and the subsequent CAS. If these two instructions were replaced with LL and SC, the latter would fail—as indeed it should—causing thread 1 to try again.

On machines with CAS, programmers must consider whether the ABA problem can arise in the algorithm at hand and, if so, take measures to avoid it. The simplest and most common technique is to devote part of each to-be-CASed word to a sequence number that is updated in pop on a successful CAS. Using this *counted pointer* technique, we can convert our stack code to the (now safe) version shown in Figure 2.7.[2]

The sequence number solution to the ABA problem requires that there be enough bits available for the number that wrap-around cannot occur in any reasonable program execution. Some machines (e.g., the x86, or the SPARC when running in 32-bit mode) provide a double-width CAS that is ideal for this purpose. If the maximum word width is required for "real" data, however, another approach may be required.

---

[2]While Treiber's technical report [Treiber, 1986] is the standard reference for the nonblocking stack algorithm, the ABA problem is mentioned as early as the 1975 edition of the System 370 manual [IBM, 1975, p. 125], and a version of the stack appears in the 1983 edition [IBM, 1983, App. A]. Treiber's personal contribution (not shown in Figure 2.7) was to observe that counted pointers are required only in the pop operation; push can safely perform a single-width CAS on the pointer alone [Michael, 2013].

```
class stack                              node* stack.pop():
    ⟨node*, int⟩ top                       repeat
void stack.push(node* n):                    ⟨o, c⟩ := top
    repeat                                   if o = null return null
        ⟨o, c⟩ := top                        n := o→next
        n→next := o                        until CAS(&top, ⟨o, c⟩, ⟨n, c+1⟩)
    until CAS(&top, ⟨o, c⟩, ⟨n, c⟩)         return o
```

Figure 2.7: The lock-free "Treiber stack," with a counted top-of-stack pointer to solve the ABA problem. It suffices to modify the count in pop only; if CAS is available in multiple widths, it may be applied to only the pointer in push.

In many programs, the programmer can reason that a given pointer will reappear in a given data structure only as a result of memory deallocation and reallocation. Note that this is *not* the case in the Treiber stack as presented here. It would be the case if we re-wrote the code to pass push a value, and had the method allocate a new node to hold it. Symmetrically, pop would deallocate the node and return the value it contained. In a garbage-collected language, deallocation will not occur so long as any thread retains a reference, so all is well. In a language with manual storage management, *hazard pointers* [Herlihy et al., 2005, Michael, 2004b] or *read–copy-update* [McKenney et al., 2001] (Section 6.3) can be used to delay deallocation until all concurrent uses of a datum have completed. In the general case (where a pointer can recur without its memory having been recycled), safe CASing may require an extra level of pointer indirection [Jayanti and Petrovic, 2003, Michael, 2004c].

2.3.2   OTHER SYNCHRONIZATION HARDWARE

Several historic machines have provided special locking instructions. The QOLB (queue on lock bit) instruction, originally designed for the Wisconsin Multicube [Goodman et al., 1989], and later adopted for the IEEE Scalable Coherent Interface (SCI) standard [Aboulenein et al., 1994], leverages a coherence protocol that maintains a linked list of copies of a given cache line. When multiple processors attempt to lock the same line at the same time, the hardware arranges to grant the requests in linked-list order. The Kendall Square KSR-1 machine [KSR, 1992] provided a similar mechanism based not on an explicit linked list, but on the implicit ordering of nodes in a ring-based network topology. As we shall see in Chapter 4, similar strategies can be emulated in software. The principal argument for the hardware approach is the ability to avoid a costly cache miss when passing the lock (and perhaps its associated data) from one processor to the next [Woest and Goodman, 1991].

The x86 allows any memory-update instruction (e.g., add or increment) to be prefixed with a special LOCK code, rendering it atomic. The benefit to the programmer is limited, however, by the fact that most instructions do not return the previous value from the modified location:

two threads executing concurrent LOCKed increments, for example, could be assured that both operations would occur, but could not tell which happened first.

Several supercomputer-class machines have provided special network hardware (generally accessed via memory-mapped I/O) for near-constant-time barrier and "Eureka" operations. These amount to cross-machine AND (all processors are ready) and OR (some processor is ready) computations. We will mention them again in Sections 5.2 and 5.3.3.

In 1991, Stone et al. proposed a multi-word variant of LL/SC, which they called "Oklahoma Update" [Stone et al., 1993] (in reference to the song "All Er Nuthin'" from the Rodgers and Hammerstein musical). Concurrently and independently, Herlihy and Moss proposed a similar mechanism for *Transactional Memory* (TM) [Herlihy and Moss, 1993]. Neither proposal was implemented at the time, but TM enjoyed a rebirth of interest about a decade later. Like queued locks, it can be implemented in software using CAS or LL/SC, but hardware implementations enjoy a substantial performance advantage [Harris et al., 2010]. As of early 2013, hardware TM has been implemented on the Azul Systems Vega 2 and 3 [Click Jr., 2009]; the experimental Sun/Oracle Rock processor [Dice et al., 2009]; the IBM Blue Gene/Q [Wang et al., 2012], zEC12 mainframe [Jacobi et al., 2012], and Power 8 processors (three independent implementations); and Intel's "Haswell" version of the x86. Additional implementations are likely to be forthcoming. We will discuss TM in Chapter 9.

Type-preserving Allocation

Both general-purpose garbage collection and hazard pointers can be used to avoid the ABA problem in applications where it might arise due to memory reallocation. Counted pointers can be used to avoid the problem in applications (like the Treiber stack) where it might arise for reasons other than memory reallocation. But counted pointers can also be used in the presence of memory reclamation. In this case, one must employ a *type-preserving* allocator, which ensures that a block of memory is reused only for an object of the same type and alignment. Suppose, for example, that we modify the Treiber stack, as suggested in the main body of the lecture, to pass push a value, and have the method allocate a new node to hold it. In this case, if a node were deallocated and reused by unrelated code (in, say, an array of floating-point numbers), it would be possible (if unlikely) that one of those numbers might match the bit pattern of a counted pointer from the memory's former life, leading the stack code to perform an erroneous operation. With a type-preserving allocator, space once occupied by a counted pointer would continue to hold such a pointer even when reallocated, and (absent wrap-around), a CAS would succeed only in the absence of reuse.

One simple implementation of a type-preserving allocator employs a Treiber stack as a free list: old nodes are pushed onto the stack when freed; new nodes are popped from the stack, or, if the stack is empty, obtained from the system memory manager. A more sophisticated implementation avoids unnecessary cache misses and contention on the top-of-stack pointer by employing a separate pool of free nodes for each thread or core. If the local pool is empty, a thread obtains a new "batch" of nodes from a backup central pool, or, if it is empty, the system memory manager. If the local pool grows too large (e.g., in a program that performs most enqueues in one thread and most dequeues in another), a thread moves a batch of nodes back to the central pool. The central pool is naturally implemented as a Treiber stack of batches.

CHAPTER 3

Essential Theory

Concurrent algorithms and synchronization techniques have a long and very rich history of formalization—far too much to even survey adequately here. Arguably the most accessible resource for practitioners is the text of Herlihy and Shavit [2008]. Deeper, more mathematical coverage can be found in the text of Schneider [1997]. On the broader topic of distributed computing (which as noted in the box on page 2 is viewed by theoreticians as a superset of shared-memory concurrency), interested readers may wish to consult the classic text of Lynch [1996].

For the purposes of the current text, we provide a brief introduction here to *safety*, *liveness*, the *consensus hierarchy*, and formal *memory models*. Safety and liveness were mentioned briefly in Section 1.4. The former says that bad things never happen; the latter says that good things eventually do. The consensus hierarchy explains the relative expressive power of hardware primitives like test_and_set (TAS) and compare_and_swap (CAS). Memory models explain which writes may be seen by which reads under which circumstances; they help to regularize the "out of order" memory references mentioned in Section 2.2.

3.1 SAFETY

Most concurrent data structures (objects) are adaptations of sequential data structures. Each of these, in turn, has its own *sequential semantics*, typically specified as a set of preconditions and postconditions for each of the methods that operate on the structure, together with *invariants* that all the methods must preserve. The sequential implementation of an object is considered safe if each method, called when its preconditions are true, terminates after a finite number of steps, having ensured the postconditions and preserved the invariants.

When designing a concurrent object, we typically wish to allow concurrent method calls ("operations"), each of which should appear to occur atomically. This goal in turn leads to at least three safety issues:

1. In a sequential program, an attempt to call a method whose precondition does not hold can often be considered an error: the program's single thread has complete control over the order in which methods are called, and can either reason that a given call is valid or else check the precondition first, explicitly, without worrying about changes between the check and the call (if $(\neg Q.\text{empty}())$ e := Q.dequeue()). In a parallel program, the potential for concurrent operation in other threads generally requires either that a method be *total* (i.e., that its precondition simply be true, allowing it to run under any circumstances), or that it use condition synchronization to wait until the precondition holds. The former option

is trivial if we are willing to return an indication that the operation is not currently valid (Q.dequeue(), for example, might return a special $\perp$ value when the queue is empty). The latter option is explored in Chapter 5.

2. Because threads may wait for one another due to locking or condition synchronization, we must address the possibility of *deadlock*, in which some set of threads are permanently waiting for each other. We consider lock-based deadlock in Section 3.1.1. Deadlocks due to condition synchronization are a matter of application-level semantics, and must be addressed on a program-by-program basis.

3. The notion of atomicity requires clarification. If operations do not actually execute one at a time in mutual exclusion, we must somehow specify the order(s) in which they are permitted to *appear* to execute. We consider several popular notions of ordering, and the differences among them, in Section 3.1.2.

3.1.1   DEADLOCK FREEDOM

As noted in Section 1.4, deadlock freedom is a safety property: it requires that there be no reachable state of the system in which some set of threads are all "waiting for one another." As originally observed by Coffman et al. [1971], deadlock requires four simultaneous conditions:

exclusive use – threads require access to some sort of non-sharable "resources"

hold and wait – threads wait for unavailable resources while continuing to hold resources they have already acquired

irrevocability – resources cannot be forcibly taken from threads that hold them

circularity – there exists a circular chain of threads in which each is holding a resource needed by the next

In shared-memory parallel programs, "non-sharable resources" often correspond to portions of a data structure, with access protected by mutual exclusion ("mutex") locks. Given that exclusive use is fundamental, deadlock can then be addressed by breaking any one of the remaining three conditions. For example:

1. We can break the hold-and-wait condition by requiring a thread that wishes to perform a given operation to request all of its locks at once. This approach is impractical in modular software, or in situations where the identities of some of the locks depend on conditions that cannot be evaluated without holding other locks (suppose, for example, that we wish to move an element atomically from set A to set $f(v)$, where v is the value of the element drawn from set A).

2. We can break the irrevocability condition by requiring a thread to release any locks it already holds when it tries to acquire a lock that is held by another thread. This approach is

commonly employed (automatically) in transactional memory systems, which are able to "back a thread out" and retry an operation (transaction) that encounters a locking conflict. It can also be used (more manually) in any system capable of dynamic *deadlock detection* (see, for example, the recent work of Koskinen and Herlihy [2008]). Retrying is complicated by the possibility that an operation may already have generated externally-visible side effects, which must be "rolled back" without compromising global invariants. We will consider roll-back further in Chapter 9.

3. We can break the circularity condition by imposing a static order on locks, and requiring that every operation acquire its locks according to that static order. This approach is slightly less onerous than requiring a thread to request all its locks at once, but still far from general. It does not, for example, provide an acceptable solution to the "move from A to $f(v)$" example in strategy 1 above.

Strategy 3 is widely used in practice. It appears, for example, in every major operating system kernel. The lack of generality, however, and the burden of defining—and respecting—a static order on locks, makes strategy 2 quite appealing, particularly when it can be automated, as it typically is in transactional memory. An intermediate alternative, sometimes used for applications whose synchronization behavior is well understood, is to consider, at each individual lock request, whether there is a feasible order in which currently active operations might complete (under worst-case assumptions about the future resources they might need in order to do so), even if the current lock is granted. The best known strategy of this sort is the *Banker's algorithm* of Dijkstra [early 1960s, 1982], originally developed for the THE operating system [Dijkstra, 1968a]. Where strategies 1 and 3 may be said to *prevent* deadlock by design, the Banker's algorithm is often described as deadlock *avoidance*, and strategy 2 as deadlock *recovery*.

3.1.2 ATOMICITY

In Section 2.2 we introduced the notion of *sequential consistency*, which requires that low-level memory accesses appear to occur in some global total order—i.e., "one at a time"—with each core's accesses appearing in program order (the order specified by the core's sequential program). When considering the order of high-level operations on a concurrent object, it is tempting to ask whether sequential consistency can help. In one sense, the answer is clearly no: correct sequential code will typically not work correctly when executed (without synchronization) by multiple threads concurrently—even on a system with sequentially consistent memory. Conversely, as we shall see in Section 3.4, one can (with appropriate synchronization) build correct high-level objects on top of a system whose memory is more relaxed.

At the same time, the notion of sequential consistency suggests a way in which we might define atomicity for a concurrent object, allowing us to infer what it means for code to be properly synchronized. After all, the memory system is a complex concurrent object from the perspective of a memory architect, who must implement load and store instructions via messages across a distributed cache-cache interconnect. Just as the designer of a sequentially consistent memory system might seek to achieve the appearance of a total order on memory accesses, consistent with per-core program order, so too might the designer of a concurrent object seek to achieve the appearance of a total order on high-level operations, consistent with the order of each thread's sequential program. In any execution that appeared to exhibit such a total order, each operation could be said to have executed atomically.

Sequential Consistency for High-Level Objects

The implementation of a concurrent object O is said to be *sequentially consistent* if, in every possible execution, the operations on O appear to occur in (have the same arguments and return values that they would have had in) some total order that is consistent with program order in each thread. Unfortunately, there is a problem with sequential consistency that limits its usefulness for high-level concurrent objects: lack of *composable orders*.

A multiprocessor memory system is, in effect, a single concurrent object, designed at one time by one architectural team. Its methods are the memory access instructions. A high-level concurrent object, by contrast, may be designed in isolation, and then used with other such objects in a single program. Suppose we have implemented object A, and have proved that in any given program, operations performed on A will appear to occur in some total order consistent with program order in each thread. Suppose we have a similar guarantee for object B. We should like to be able to guarantee that in any given program, operations on A and B will appear to occur in some *single* total order consistent with program order in each thread. That is, we should like the implementations of A and B to *compose*. Sadly, they may not.

As a simple if somewhat contrived example, consider a replicated integer object, in which threads read their local copy (without synchronization) and update all copies under protection of a lock:

```
// initially L is free and A[i] = 0 ∀ i ∈ T

void put(int v):                          int get():
    L.acquire()                               return A[self]
    for i ∈ T
        A[i] := v
    L.release()
```

(Throughout this lecture, we use T to represent the set of thread ids. For the sake of convenience, we assume that the set is sufficiently dense that we can use it to index arrays.)

Because of the lock, put operations are totally ordered. Further, because a get operation performs only a single (atomic) access to memory, it is easily ordered with respect to all puts—after those that have updated the relevant element of A, and before those that have not. It is straightforward to identify a total order on operations that respects these constraints and that is consistent with program order in each thread. In other words, our counter is sequentially consistent.

On the other hand, consider what happens if we have *two* counters—call them X and Y. Because get operations can occur "in the middle of" a put at the implementation level, we can imagine a scenario in which threads $T3$ and $T4$ perform gets on X and Y while both objects are being updated—and see the updates in opposite orders:

| | local values of shared objects | | | |
| | T1 | T2 | T3 | T4 |
	X Y	X Y	X Y	X Y
initially	0 0	0 0	0 0	0 0
$T1$ begins X.put(1)	1 0	1 0	0 0	0 0
$T2$ begins Y.put(1)	1 1	1 1	0 1	0 0
$T3$: X.get() returns 0	1 1	1 1	0 1	0 0
$T3$: Y.get() returns 1	1 1	1 1	0 1	0 0
$T1$ finishes X.put(1)	1 1	1 1	1 1	1 0
$T4$: X.get() returns 1	1 1	1 1	1 1	1 0
$T4$: Y.get() returns 0	1 1	1 1	1 1	1 0
$T2$ finishes Y.put(1)	1 1	1 1	1 1	1 1

At this point, the put to Y has happened before the put to X from $T3$'s perspective, but after the put to X from $T4$'s perspective. To solve this problem, we might require the implementation of a shared object to ensure that updates appear to other threads to happen at some single point in time.

But this is not enough. Consider a software emulation of the hardware *write buffers* described in Section 2.2.1. To perform a put on object X, thread T inserts the desired new value into a local queue and continues execution. Periodically, a helper thread drains the queue and applies the updates to the master copy of X, which resides in some global location. To perform a get, T inspects the local queue (synchronizing with the helper as necessary) and returns any pending update; otherwise it returns the global value of X. From the point of view of every thread other than T, the update occurs when it is applied to the global value of X. From T's perspective, however, it happens early, and, in a system with more than one object, we can easily obtain the "bow tie" causality loop of Figure 2.3. This scenario suggests that we require updates to appear to other threads at the same time they appear to the updater—or at least before the updater continues execution.

Linearizability

To address the problem of composability, Herlihy and Wing introduced the notion of *linearizability* [1990]. For more than 20 years it has served as the standard ordering criterion for high-level concurrent objects. The implementation of object O is said to be linearizable if, in every possible execution, the operations on O appear to occur in some total order that is consistent not only with program order in each thread but also with any ordering that threads are able to observe by other means.

More specifically, linearizability requires that each operation appear to occur instantaneously at some point in time between its call and return. The "instantaneously" part of this requirement precludes the shared counter scenario above, in which $T3$ and $T4$ have different views of partial updates. The "between its call and return" part of the requirement precludes the software write buffer scenario, in which a put by thread T may not be visible to other threads until after it has returned.

For the sake of precision, it should be noted that there is no absolute notion of objective time in a parallel system, any more than there is in Einsteinian physics. (For more on the notion of time in parallel systems, see the classic paper by Lamport [1978].) What really matters is observable orderings. When we say that an event must occur at a single instant in time, what we mean is that it must be impossible for thread A to observe that an event has occurred, for A to subsequently communicate with thread B (e.g., by writing a variable that B reads), and then for B to observe that the event has not yet occurred.

To help us reason about the linearizability of a concurrent object, we typically identify a *linearization point* within each method at which a call to that method can be said to have occurred. If we choose these points properly, then whenever the linearization point of operation A precedes the linearization point of operation B, we will know that operation A, as a whole, linearizes before operation B.

In the trivial case in which every method is bracketed by the acquisition and release of a common object lock, the linearization point can be anywhere inside the method—we might arbitrarily place it at the lock release. In an algorithm based on fine-grain locks, the linearization point might correspond to the release of some particular one of the locks.

In nonblocking algorithms, it is common to associate linearization with a specific instruction (a load, store, or other atomic primitive) and then argue that any implementation-level memory updates that are visible before the linearization point will be recognized by other threads as merely preparation, and any that can be seen to occur after it will be recognized as merely cleanup. In the nonblocking stack of Figure 2.7, a successful push or pop can be said to linearize at its final CAS instruction; an unsuccessful pop (one that returns null) can be said to linearize at the load of top.

In a complex method, we may need to identify multiple possible linearization points, to accommodate branching control flow. In other cases, the outcome of tests at run time may allow us to argue that a method linearized at some point *earlier* in its execution (an example of this sort can be found in Section 8.3). There are even algorithms in which the linearization point of a method is determined by behavior in some *other* thread. All that really matters is that there be a total order on the linearization points, and that the behavior of operations, when considered in that order, be consistent with the object's sequential semantics.

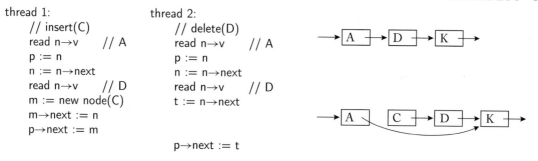

```
thread 1:                          thread 2:
    // insert(C)                       // delete(D)
    read n→v      // A                 read n→v      // A
    p := n                             p := n
    n := n→next                        n := n→next
    read n→v      // D                 read n→v      // D
    m := new node(C)                   t := n→next
    m→next := n
    p→next := m
                                       p→next := t
```

Figure 3.1: Dynamic trace of improperly synchronized list updates. This execution can lose node C even on a sequentially consistent machine.

Given linearizable implementations of objects A and B, one can prove that in every possible program execution, the operations on A and B will appear to occur in some *single* total order that is consistent both with program order in each thread and with any other ordering that threads are able to observe. In other words, linearizable implementations of concurrent objects are composable. Linearizability is therefore sometimes said to be a *local* property [Herlihy and Wing, 1990, Weihl, 1989]: the linearizability of a system as a whole depends only on the (local) linearizability of its parts.

Hand-over-hand Locking (Lock Coupling).   As an example of linearizability achieved through fine-grain locking, consider the task of parallelizing a set abstraction implemented as a sorted, singly-linked list with insert, remove, and lookup operations. Absent synchronization, it is easy to see how the list could become corrupted. In Figure 3.1, the code at left shows a possible sequence of statements executed by thread 1 in the process of inserting a new node containing the value C, and a concurrent sequence of statements executed by thread 2 in the process of deleting the node containing the value D. If interleaved as shown (with thread 1 performing its last statement between thread 2's last two statements), these two sequences will transform the list at the upper right into the non-list at the lower right, in which the node containing C has been lost.

Clearly a global lock—forcing either thread 1 or thread 2 to complete before the other starts—would linearize the updates and avoid the loss of C. It can be shown, however, that linearizability can also be maintained with a fine-grain locking protocol in which each thread holds at most two locks at a time, on adjacent nodes in the list [Bayer and Schkolnick, 1977]. By retaining the right-hand lock while releasing the left-hand and then acquiring the right-hand's successor, a thread ensures that it is never overtaken by another thread during its traversal of the list. In Figure 3.1, thread 1 would hold locks on the nodes containing A and D until done inserting the node containing C. Thread 2 would need these same two locks before reading the content of the node containing A. While threads 1 and 2 cannot make their updates simultaneously, one can "chase the other down the list" to the point where the updates are needed, achieving substantially higher concurrency than is possible with a global lock. Similar "hand-over-hand" locking techniques are commonly used in concurrent trees and other pointer-based data structures.

Serializability

Recall that the purpose of an ordering criterion is to clarify the meaning of atomicity. By requiring an operation to complete at a single point in time, and to be visible to all other threads before it returns to its caller, linearizability guarantees that the order of operations on any given concurrent object will be consistent with all other observable orderings in an execution, including those of other concurrent objects.

The flip side of this guarantee is that the linearizability of individual operations does not necessarily imply linearizability for operations that manipulate more than one object, but are still intended to execute as a single atomic unit.

Consider a banking system in which thread 1 transfers $100 from account A to account B, while thread 2 adds the amounts in the two accounts:

```
// initially A.balance() = B.balance() = 500

thread 1:                              thread 2:
    A.withdraw(100)
                                           sum := A.balance()     // 400
                                           sum += B.balance()     // 900

    B.deposit(100)
```

If we think of A and B as separate objects, then the execution can linearize as suggested by vertical position on the page, but thread 2 will see a cross-account total that is $100 "too low." If we wish to treat the code in each thread as a single atomic unit, we must disallow this execution—something that neither A nor B can do on its own. We need, in short, to be able to *combine* smaller atomic operations into larger ones—not just perform the smaller ones in a mutually consistent order. Where linearizability ensures that the orders of separate objects will compose "for free," multi-object atomic operations will generally require some sort of global or distributed control.

Multi-object atomic operations are the hallmark of database systems, which refer to them as *transactions*. Transactional memory (the subject of Chapter 9) adapts transactions to shared-memory parallel computing, allowing the programmer to request that a multi-object operation like thread 1's transfer or thread 2's sum should execute atomically.

The simplest ordering criterion for transactions—both database and memory—is known as *serializability*. Transactions are said to serialize if they have the same effect they would have had if executed one at a time in some total order. For transactional memory (and sometimes for databases as well), we can extend the model to allow a thread to perform a series of transactions, and require that the global order be consistent with program order in each thread.

It turns out to be NP-hard to determine whether a given set of transactions (with the given inputs and outputs) is serializable [Papadimitriou, 1979]. Fortunately, we seldom need to make such a determination in practice. Generally all we really want is to ensure that the current execution will be serializable—something we can achieve with conservative (sufficient but not necessary) measures. A global lock is a trivial solution, but admits no concurrency. Databases and most TM systems employ more elaborate fine-grain locking. A few TM systems employ nonblocking techniques.

If we regard the objects to be accessed by a transaction as "resources" and revisit the conditions for deadlock outlined at the beginning of Section 3.1.1, we quickly realize that a transaction may, in the general case, need to access some resources before it knows which others it will need. Any implementation of serializability based on fine-grain locks will thus entail not only "exclusive use," but also both "hold and wait" and "circularity." To address the possibility of deadlock, a database or lock-based TM system must be prepared to break the "irrevocability" condition by releasing locks, rolling back, and retrying conflicting transactions.

Like branch prediction or CAS-based fetch_and_Φ, this strategy of proceeding "in the hope" that things will work out (and recovering when they don't) is an example of *speculation*. So-called *lazy* TM systems take this even further, allowing conflicting (non-serializable) transactions to proceed in parallel until one of them is ready to *commit*—and only then *aborting* and rolling back the others.

Two-Phase Locking.   As an example of fine-grain locking for serializability, consider a simple scenario in which transactions 1 and 2 read and update symmetric variables:

```
// initially x = y = 0
transaction 1:              transaction 2:
    t1 := x                     t2 := y
    y++                         x++
```

Left unsynchronized, this code could result in t1 = t2 = 0, even on a sequentially consistent machine—something that should not be possible if the transactions are to serialize. A global lock would solve the problem, but would be far too conservative for transactions larger than the trivial ones shown here. If we associate fine-grain locks with individual variables, we still run into trouble if thread 1 releases its lock on x before acquiring the lock on y, and thread 2 releases its lock on y before acquiring the lock on x.

It turns out [Eswaran et al., 1976] that serializability can always be guaranteed if threads acquire all their locks (in an "expansion phase") before releasing any of them (in a "contraction phase"). As we have observed, this *two-phase locking* convention admits the possibility of deadlock: in our example, transaction 1 might lock x and transaction 2 lock y before either attempts to acquire the other. To detect the problem and trigger rollback, a system based on two-phase locking may construct and maintain a dependence graph at run time. Alternatively (and more conservatively), it may simply limit the time it is willing to wait for locks, and assume the worst when this timeout is exceeded.

Strict Serializability

The astute reader may have noticed the strong similarity between the definitions of sequential consistency (for high-level objects) and serializability (with the extension that allows a single thread to perform a series of transactions). The difference is simply that transactions need to be able to access a dynamically chosen set of objects, while sequential consistency is limited to a predefined set of single-object operations.

The similarity between sequential consistency and serializability leads to a common weakness: the lack of required consistency with other orders that may be observed by a thread. It was by requiring such "real-time" ordering that we obtained composable orders for single-object operations in the definition of linearizability. Real-time ordering is also important for its own sake in many applications. Without it we might, for example, make a large deposit to a friend's bank account, tell the person we had done so, and yet still encounter an "insufficient funds" message in response to a (subsequent!) withdrawal request. To avoid such arguably undesirable scenarios, many database systems—and most TM systems—require *strict serializability*, which is simply ordinary serializability augmented with real-time order: transactions are said to be strictly serializable if they have the same effect they would have had if executed one at a time in some total order that is consistent with program order (if any) in each thread, and with any other order the threads may be able to observe. In particular, if transaction A finishes before transaction B begins, then A must appear before B in the total order. As it turns out, two-phase locking suffices to ensure strict serializability, but certain other implementations of "plain" serializability do not.

Relationships Among the Ordering Criteria

Table 3.1 summarizes the relationships among sequential consistency (for high-level objects), linearizability, serializability, and strict serializability. A system that correctly implements any of these four criteria will provide the appearance of a total order on operations, consistent with per-thread program order. Linearizability and strict serializability add consistency with "real-time" order. Serializability and strict serializability add the ability to define multi-object atomic operations. Of the four criteria, only linearizability is local: it guarantees that operations on separate objects always occur in a mutually consistent order, and it declines, as it were, to address multi-object operations.

To avoid confusion, it should be noted that we have been using the term "composability" to mean that we can merge (compose) the orders of operations on separate objects into a single mutually consistent order. In the database and TM communities, "composability" means that we can combine (compose) individual atomic operations into larger, still atomic (i.e., serializable) operations. We will return to this second notion of composability in Chapter 9. It is straightforward to provide in a system based on speculation; it is invariably supported by databases and transactional memory. It cannot be supported, in the general case, by conservative locking strategies. Somewhat ironically, linearizability might be said to facilitate composable orders by disallowing composable operations.

Table 3.1: Properties of standard ordering criteria

SC = sequential consistency; L = linearizability;
S = serializability; SS = strict serializability.

	SC	L	S	SS
Equivalent to a sequential order	+	+	+	+
Respects program order in each thread	+	+	+	+
Consistent with other ordering ("real time")	−	+	−	+
Can touch multiple objects atomically	−	−	+	+
Local: reasoning based on individual objects only	−	+	−	−

3.2 LIVENESS

Safety properties—the subject of the previous section—ensure that bad things never happen: threads are never deadlocked; atomicity is never violated; invariants are never broken. To say that code is correct, however, we generally want more: we want to ensure forward progress. Just as we generally want to know that a sequential program will produce a correct answer eventually (not just fail to produce an incorrect answer), we generally want to know that invocations of concurrent operations will complete their work and return.

An object method is said to be *blocking* (in the theoretical sense described in the box on page 7) if there is some reachable state of the system in which a thread that has called the method will be unable to return until some other thread takes action. Lock-based algorithms are inherently blocking: a thread that holds a lock precludes progress on the part of any other thread that needs the same lock. Liveness proofs for lock-based algorithms require not only that the code be deadlock-free, but also that critical sections be free of infinite loops, and that all threads continue to execute.

A method is said to be *nonblocking* if there is no reachable state of the system in which an invocation of the method will be unable to complete its execution and return. Nonblocking algorithms have the desirable property that inopportune preemption (e.g., of a lock holder) never precludes forward progress in other threads. In some environments (e.g., a system with high fault-tolerance requirements), nonblocking algorithms may also allow the system to survive when a thread crashes or is prematurely killed. We consider several variants of nonblocking progress in Section 3.2.1.

In both blocking and nonblocking algorithms, we may also care about *fairness*—the relative rates of progress of different threads. We consider this topic briefly in Section 3.2.2.

3.2.1 NONBLOCKING PROGRESS

Given the difficulty of guaranteeing any particular rate of execution (in the presence of timesharing, cache misses, page faults, and other sources of variability), we generally speak of progress in terms of abstract program *steps* rather than absolute time.

A method is said to be *wait free* (the strongest variant of nonblocking progress) if it is guaranteed to complete in some bounded number of its own program steps. (This bound need not be statically known.) A method M is said to be *lock free* (a somewhat weaker variant) if *some* thread is guaranteed to make progress (complete an operation on the same object) in some bounded number of M's program steps. M is said to be *obstruction free* (the weakest variant of nonblocking progress) if it is guaranteed to complete in some bounded number of program steps if no other thread executes any steps during that same interval.

Wait freedom is sometimes referred to as *starvation freedom*: a given thread is never prevented from making progress. Lock freedom is sometimes referred to as *livelock freedom*: an individual thread may starve, but the system as a whole is never prevented from making forward progress (equivalently: no set of threads can actively prevent each other from making progress indefinitely). Obstruction-free algorithms can suffer not only from starvation but also from livelock; if all threads but one "hold still" long enough, however, the one running thread is guaranteed to make progress.

Many practical algorithms are lock free or obstruction free. Treiber's stack, for example (Section 2.3.1), is lock-free, as is the widely used queue of Michael and Scott (Section 8.2). Obstruction freedom was first described in the context of Herlihy et al.'s double-ended queue [2003a] (Section 8.6.2). It is also provided by several TM systems (among them the DSTM of Herlihy et al. [2003b], the ASTM of Marathe et al. [2005], and the work of Marathe and Moir [2008]). Moir and Shavit [2005] provide an excellent survey of concurrent data structures, including coverage of nonblocking progress. Sundell and Tsigas [2008a] describe a library of nonblocking data structures.

Wait-free algorithms are significantly less common. Herlihy [1991] demonstrated that any sequential data structure can be transformed, automatically, into a wait-free concurrent version, but the construction is highly inefficient. Recent work by Kogan and Petrank [2012] (building on a series of intermediate results) has shown how to reduce the time overhead dramatically, though space overhead remains proportional to the maximum number of threads in the system.

Most wait-free algorithms—and many lock-free algorithms—employ some variant of *helping*, in which a thread that has begun but not completed its operation may be assisted by other threads, which need to "get it out of the way" so they can perform their own operations without an unbounded wait. Other lock-free algorithms—and even a few wait-free algorithms—are able to make do without helping. As a simple example, consider the following implementation of a wait-free increment-only counter:

```
initially C[i] = 0 ∀ i ∈ 𝒯

void inc():                    int val():
    C[self]++                      rtn := 0
                                   for i in [1..N]
                                       rtn +:= C[i]
                                   return rtn
```

Here the aggregate counter value is taken to be the sum of a set of per-thread values. Because each per-thread value is monotonically increasing (and always by exactly one), so is the aggregate sum. Given this observation, one can prove that the value returned by the val method will have been correct at some point between its call and its return: it will be bounded above by the number of inc operations that were called before val returned, and below by the number that returned before it was called. In other words, val is linearizable, though its linearization point cannot in general be statically determined. Because both inc and val comprise a bounded (in this case, statically bounded) number of program steps, the methods are wait free.

3.2.2   FAIRNESS

Obstruction freedom and lock freedom clearly admit behavior that defies any notion of fairness: both allow an individual thread to take an unbounded number of steps without completing an operation. Even wait freedom allows an operation to execute an arbitrary number of steps (helping or deferring to peers) before completing, so long as the number is bounded in any given situation.

We shall often want stronger guarantees. In a wait-free algorithm, we might hope for a static bound, across all invocations, on the number of steps required to complete an operation. In a blocking algorithm, we might hope for a bound on the number of competing operations that may complete before a given thread makes progress. If threads repeatedly invoke a certain set of operations, we might even wish to bound the ratio of their "success" rates. These are only a few of the possible ways in which "fairness" might be defined. Without dwelling on particular definitions, we will consider algorithms in subsequent chapters whose behavior ranges from potentially very highly skewed (e.g., test_and_set locks that avoid starvation only when there are periodic quiescent intervals, when the lock is free and no thread wants it), to strictly first-come, first-served (e.g., locks in which a thread employs a wait-free protocol to join a FIFO queue). We will also consider intermediate options, such as locks that deliberately balance locality (for performance) against uniformity of service to threads.

In any practical system, forward progress relies on the assumption that any continually unblocked thread will eventually execute another program step. Without such minimal fairness within the implementation, a system could be "correct" without doing anything at all! Significantly, even this minimal fairness depends on scheduling decisions at multiple system levels—in the hardware, the operating system, and the language runtime—all of which ensure that runnable threads continue to run.

When threads may block for mutual exclusion or condition synchronization, we shall in most cases want to insist that the system display what is known as *weak fairness*. This property guarantees that any thread waiting for a condition that is continuously true (or a lock that is continuously available) eventually executes another program step. Without such a guarantee, program behavior may be highly unappealing. Imagine a web server, for example, that never accepts requests from a certain client connection if requests are available from any other client.

In the following program fragment, weak fairness precludes an execution in which thread 1 spins forever: thread 2 must eventually notice that f is false, complete its wait, and set f to true, after which thread 1 must notice the change to f and complete:

```
initially f = false
thread 1:                          thread 2:
    await f                            await ¬f
                                       f := true
```

Here we have used the notation await (*condition*) as shorthand for

```
while ¬condition
    // spin
fence(R‖RW)
```

Many more stringent definitions of fairness are possible. In particular, *strong fairness* requires that any thread waiting for a condition that is true infinitely often (or a lock that is available infinitely often) eventually executes another program step. In the following program fragment, for example, weak fairness admits an execution in which thread 1 spins forever, but strong fairness requires thread 2 to notice one of the "windows" in which g is true, complete its wait, and set f to true, after which thread 1 must notice the change and complete:

```
initially f = g = false
thread 1:                          thread 2:
    while ¬f                           await (g)
        g := true                      f := true
        g := false
```

Strong fairness is difficult to truly achieve: it may, for example, require a scheduler to recheck every awaited condition whenever one of its constituent variables is changed, to make sure that any thread at risk of starving is given a chance to run. Any deterministic strategy that considers only a subset of the waiting threads on each state change risks the possibility of deterministically ignoring some unfortunate thread every time it is able to run.

Fortunately, statistical "guarantees" typically suffice in practice. By considering a randomly chosen thread—instead of all threads—when a scheduling decision is required, we can drive the probability of starvation arbitrarily low. A truly random choice is difficult, of course, but various pseudorandom approaches appear to work quite well. At the hardware level, interconnects and coherence protocols are designed to make it unlikely that a "race" between two cores (e.g., when performing near-simultaneous CAS instructions on a previously uncached location) will always be resolved the same way. Within the operating system, runtime, or language implementation, one can "randomize" the interval between checks of a condition using a pseudorandom number generator or even the natural "jitter" in execution time of nontrivial instruction sequences on complex modern cores.

Weak and strong fairness address worst-case behavior, and allow executions that still seem grossly unfair from an intuitive perspective (e.g., executions in which one thread succeeds a million times more often than another). Statistical "randomization," by contrast, may achieve intuitively very fair behavior without absolutely precluding worst-case starvation.

Much of the theoretical groundwork for fairness was laid by Nissim Francez [1986]. Proofs of fairness are typically based on *temporal logic*, which provides operators for concepts like "always" and "eventually." A brief introduction to these topics can be found in the text of Ben-Ari [2006, Chap. 4]; much more extensive coverage can be found in Schneider's comprehensive work on the theory of concurrency [1997].

3.3   THE CONSENSUS HIERARCHY

In Section 2.3 we noted that CAS and LL/SC are *universal* atomic primitives—capable of implementing arbitrary single-word fetch_and_Φ operations. We suggested—implicitly, at least—that they are fundamentally more powerful than simpler primitives like TAS, Swap, FAI, and FAA. Herlihy formalized this notion of relative power in his work on wait-free synchronization [1991], previously mentioned in Section 3.2.1. The formalization is based on the classic *consensus problem*.

Originally formalized by Fischer, Lynch, and Paterson [1985] in a distributed setting, the consensus problem involves a set of potentially unreliable threads, each of which "proposes" a value. The goal is for the reliable threads to agree on one of the proposed values—a task the authors proved to be impossible with asynchronous messages. Herlihy adapted the problem to the shared-memory setting, where powerful atomic primitives can circumvent impossibility. Specifically, Herlihy suggested that such primitives (or, more precisely, the objects on which those primitives operate) be classified according the number of threads for which they can achieve *wait-free* consensus.

It is easy to see that an object with a TAS method can achieve wait-free consensus for two threads:

```
// initially L = 0; proposal[0] and proposal[1] are immaterial
agree(i):
    proposal[self].store(i)
    if TAS(L) return i
    else return proposal[1−self].load()
```

Herlihy was able to show that this is the best one can do: TAS objects (even an arbitrary number of them) cannot achieve wait-free consensus for more than two threads. Moreover ordinary loads and stores cannot achieve wait-free consensus at all—even for only two threads. An object supporting CAS, on the other hand (or equivalently LL/SC), can achieve wait-free consensus for an arbitrary number of threads:

```
// initially v = ⊥
agree(i):
      if CAS(&v, ⊥, i) return i
      else return v
```

One can, in fact, define an infinite hierarchy of atomic objects, where those appearing at level k can achieve wait-free consensus for k threads but no more. Objects supporting CAS or LL/SC are said to have *consensus number* ∞. Objects with other common primitives—including TAS, swap, FAI, and FAA—have consensus number 2. One can define atomic objects at intermediate levels of the hierarchy, but these are not typically encountered on real hardware.

3.4   MEMORY MODELS

As described in Section 2.2, most modern multicore systems are *coherent* but not *sequentially consistent*: changes to a given variable are serialized, and eventually propagate to all cores, but accesses to different locations may appear to occur in different orders from the perspective of different threads—even to the point of introducing apparent causality loops. For programmers to reason about such a system, we need a *memory model*—a formal characterization of its behavior. Such a model can be provided at the hardware level—SPARC TSO is one example—but when programmers write code in higher-level languages, they need a language-level model. Just as a hardware-level memory model helps the compiler writer or library builder determine where to employ special ordering instructions, a language-level model helps the application programmer determine where to employ synchronization operations or atomic variable accesses.

There is an extensive literature on language-level memory models. Good starting points include the tutorial of Adve and Gharachorloo [1996], the lecture of Sorin et al. [2011], and the articles introducing the models of Java [Manson et al., 2005] and C++ [Boehm and Adve, 2008]. Details vary considerably from one model to another, but most now share a similar framework.

Consensus and Mutual Exclusion

A solution to the consensus problem clearly suffices for "one-shot" mutual exclusion (a.k.a. *leader election*): each thread proposes its own id, and the agreed-upon value indicates which thread is able to enter the critical section. Consensus is not *necessary* however: the winning thread needs to know that it can enter the critical section, but other threads only need to know that they have lost—they don't need to know who won. TAS thus suffices to build a wait-free *try lock* (one whose acquire method returns immediately with a success or failure result) for an arbitrary number of threads.

It is tempting to suggest that one might solve the consensus problem using mutual exclusion, by having the winner of the competition for the lock write its id into a location visible to the losers. This approach, however, cannot be wait free: it fails to bound the number of steps required by a losing thread. If the winner acquires the lock and then pauses—or dies—before writing down its id, the losers may execute their spin loops an arbitrary number of times. This is the beauty of CAS or LL/SC: it allows a thread to win the competition *and* write down its value in a single indivisible step.

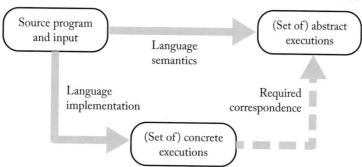

Figure 3.2: Program executions, semantics, and implementations. A valid implementation must produce only those concrete executions whose output agrees with that of some abstract execution allowed by language semantics for the given program and input.

3.4.1   FORMAL FRAMEWORK

Informally, a memory model specifies which values (i.e., which writes) may be seen by any given read. At the language level, more precise definitions depend on the notion of an *abstract program execution*.

Programming language semantics are typically defined in terms of execution on some abstract machine, with language-appropriate built-in types, control-flow constructs, etc. For a given source program and input, an abstract execution is a set of sequences, one per thread, of reads, writes, and other atomic program steps, each of which inspects and/or changes the state of the abstract machine (i.e., memory). For any given program and input, certain executions of the abstract machine are said to be *valid* according to the language's semantics. Language semantics can be seen, constructively, as a mapping from programs and inputs to (possibly infinite) sets of valid executions. The semantics can also be seen, nonconstructively, as a predicate: given a program, its input, and an abstract execution, is the execution valid?

For the sake of convenience, the remainder of this section adopts the nonconstructive point of view, at least with respect to the memory model. In practice, it is the language implementor who needs to construct an execution—a *concrete* execution—for a given program and input. A language implementation is *safe* if for every well formed source program and input, every concrete execution it produces corresponds to (produces the same output as) some valid abstract execution of that program on that input (Figure 3.2). The implementation is *live* if produces at least one such concrete execution whenever the set of abstract executions is nonempty. To show that a language implementation is safe (a task beyond the scope of this lecture), one can define a mapping from concrete to abstract executions, and then demonstrate that all the resulting abstract executions will be valid.

At the programming language level, a memory model is the portion of language semantics that determines whether the values read from variables in an abstract execution are valid, given the values written by other program steps. In a single-threaded program, the memory model is trivial: there is a total order on program steps in any given execution, and the value read in a given step is valid if and only if it matches the value written by the most recent prior write to the same variable—or the initial value if there is no such write.

In a multithreaded program, the memory model is substantially more complex, because a read in one thread may see a value written by a write in a different thread. An execution is said to be sequentially consistent (in the sense of Section 2.2) if there exists a total order on memory operations (across all threads) that explains all values read. As we have seen, of course, most hardware is not sequentially consistent. Absent language restrictions, a concrete execution is unlikely to be both fast and sequentially consistent. For the sake of efficiency, the memory model for a concurrent language therefore typically requires only a partial order—known as *happens-before*—on the steps of an abstract execution. Using this partial order, the model then defines a *writes-seen* relation that identifies, for every read, the writes whose values may be seen.

To define the happens-before order, most memory models begin by distinguishing between "ordinary" and "synchronizing" steps in the abstract execution. Ordinary steps are typically reads and writes of scalar variables. Depending on the language, synchronizing steps might be lock acquire and release, transactions, monitor entry and exit, message send and receive, or reads and writes of special atomic variables. Like ordinary steps, most synchronizing steps read or write values in memory (an acquire operation, for example, changes a lock from "free" to "held"). A sequentially consistent execution must explain these reads and writes, just as it does those effected by accesses to ordinary scalar variables.

Given the definition of ordinary and synchronizing steps, we proceed incrementally as follows:

Program order is the union of a collection of disjoint total orders, each of which captures the steps performed by one of the program's threads. Each thread's steps must be allowable under the language's sequential semantics, given the values returned by read operations.

Synchronization order is a total order, across all threads, on all synchronizing steps. This order must be consistent with program order within each thread. It must also explain the values read and written by the synchronizing steps (this will ensure, for example, that acquire and release operations on any given lock occur in alternating order). Crucially, synchronization order is not specified by the source program. An execution is valid only if there exists a synchronization order that leads, as described below, to a writes-seen relation that explains the values read by ordinary steps.

Synchronizes-with order is a subset of synchronization order induced by language semantics. In a language based on transactional memory, the subset may be trivial: all transactions are globally ordered. In a language based on locks, each release operation may synchronize with the next acquire of the same lock in synchronization order, but other synchronizing steps may be unordered.

Happens-before order is the transitive closure of program order and synchronizes-with order. It captures all the ordering the language guarantees.

To complete a memory model, these order definitions must be augmented with a writes-seen relation. To understand such relations, we first must understand the notion of a *data race*.

3.4.2 DATA RACES

Language semantics specify classes of ordinary program steps that *conflict* with one another. A write, for example, is invariably defined to conflict with either a read or a write to the same variable. A program is said to have a data race if, for some input, it has a sequentially consistent execution in which two conflicting ordinary steps, performed by different threads, are adjacent in the total order. Data races are problematic because we don't normally expect an implementation to force ordinary steps in different threads to occur in a particular order. If an implementation yields (a concrete execution corresponding to) an abstract execution in which the conflicting steps occur in one order, we need to allow it to yield another abstract execution (of the same program on the same input) in which all prior steps are the same but the conflicting steps are reversed. It is easy to construct examples (e.g., as suggested in Figure 2.3) in which the remainder of this second execution cannot be sequentially consistent.

Given the definitions in Section 3.4.1, we can also say that an abstract execution has a data race if it contains a pair of conflicting steps that are not ordered by happens-before. A program then has a data race if, for some input, it has an execution containing a data race. This definition turns out to be equivalent to the one based on sequentially consistent executions. The key to the equivalence is the observation that arcs in the synchronizes-with order, which contribute to happens-before, correspond to ordering constraints in sequentially consistent executions. The same language rules that induce a synchronizes-with arc from, say, the release of a lock to the

---

Synchronization Races

The definition of a data race is designed to capture cases in which program behavior may depend on the order in which two ordinary accesses occur, and this order is not constrained by synchronization. In a similar fashion, we may wish to consider cases in which program behavior depends on the outcome of synchronization operations.

For each form of synchronization operation, we can define a notion of conflict. Acquire operations on the same lock, for example, conflict with one another, while an acquire and a release do not—nor do operations on different locks. A program is said to have a *synchronization race* if it has two sequentially consistent executions with a common prefix, and the first steps that differ are conflicting synchronization operations. Together, data races and synchronization races constitute the class of *general races* [Netzer and Miller, 1992].

Because we assume the existence of a total order on synchronizing steps, synchronization races never compromise sequential consistency. Rather, they provide the means of controlling and exploiting nondeterminism in parallel programs. In any case where we wish to allow conflicting high-level operations to occur in arbitrary order, we design a synchronization race into the program to mediate the conflict.

---

subsequent acquire also force the release to appear before the acquire in the total order of any sequentially consistent execution.

In an execution without any data races, the writes-seen relation is straightforward: the lack of unordered conflicting accesses implies that all reads and writes of a given location are ordered by happens-before. Each read can then return the value written by the (unique) most recent prior write of the same location in happens-before order—or the initial value if there is no such write. More formally, one can prove that all executions of a data-race-free program are sequentially consistent: any total order consistent with happens-before will explain the program's reads. Moreover, since our (first) definition of a data race was based only on sequentially consistent executions, we can provide the programmer with a set of rules that, if followed, will always lead to sequentially consistent executions, with no need to reason about possible relaxed behavior of the underlying hardware. Such a set of rules is said to constitute a *programmer-centric* memory model [Adve and Hill, 1990].

In effect, a programmer-centric model is a contract between the programmer and the implementation: if the programmer follows the rules (i.e., write data-race-free programs), the implementation will provide the illusion of sequential consistency. Moreover, given the absence of races, any region of code that contains no synchronization (and that does not interact with the "outside world" via I/O or syscalls) can be thought of as atomic: it cannot—by construction—interact with other threads.

But what about programs that *do* have data races? Some researchers have argued that such programs are simply buggy, and should have undefined behavior. This is the approach adopted by C++ [Boehm and Adve, 2008] and, subsequently, C. It rules out certain categories of programs (e.g., chaotic relaxation [Chazan and Miranker, 1969]), but the language designers had little in the way of alternatives: in the absence of type safety it is nearly impossible to limit the potential impact of a data race. The resulting model is quite simple (at least in the absence of variables that have been declared atomic): if a C or C++ program has a data race on a given input, its behavior is undefined; otherwise, it follows one of its sequentially consistent executions.

Unfortunately, in a language like Java, even buggy programs need to have well defined behavior, to safeguard the integrity of the virtual machine (which may be embedded in some larger, untrusting system). The obvious approach is to say that a read may see the value written by the most recent write on any backward path through the happens-before graph, or by any *incomparable* write (one that is unordered with respect to the read). Unfortunately, as described by Manson et al. [2005], this approach is overly restrictive: it precludes the use of several important compiler optimizations. The actual Java model defines a notion of "incremental justification" that may allow a read to see a value that might have been written by an incomparable write in some *other* hypothetical execution. The details are surprisingly subtle and complex, and as of 2012 it is still unclear whether the current specification is correct, or could be made so.

3.4.3   REAL-WORLD MODELS

As of this writing, Java and C/C++ are the only widely used parallel programming languages whose definitions attempt to precisely specify a memory model. Ada [Ichbiah et al., 1991] was the first language to introduce an explicitly relaxed (if informally specified) memory model. It was designed to facilitate implementation on both shared-memory and distributed hardware: variables shared between threads were required to be consistent only in the wake of explicit message passing (rendezvous). The reference implementations of several scripting languages (notably Ruby and Python) are sequentially consistent, though other implementations [JRuby, Jython] are not.

A group including representatives of Intel, Oracle, IBM, and Red Hat has proposed transactional extensions to C++ [Adl-Tabatabai et al., 2012]. In this proposal, begin_transaction and end_transaction markers contribute to the happens-before order inherited from standard C++. So-called *relaxed* transactions are permitted to contain other synchronization operations (e.g., lock acquire and release); *atomic* transactions are not. Dalessandro et al. [2010b] have proposed an alternative model in which atomic blocks are fundamental, and other synchronization mechanisms (e.g., locks) are built on top of them.

If we wish to allow programmers to create new synchronization mechanisms or nonblocking data structures (and indeed if any of the built-in synchronization mechanisms are to be written in high-level code, rather than assembler), then the memory model must define synchronizing steps that are more primitive than lock acquire and release. Java allows a variable to be labeled volatile, in which case reads and writes that access it are included in the global synchronization order, with each read inducing a synchronizes-with arc (and thus a happens-before arc) from the (unique) preceding write to the same location. C and C++ provide a substantially more complex facility, in which variables are labeled atomic, and an individual read, write, or fetch_and_Φ operation can be labeled as an acquire access, a release access, both, or neither. By default, operations on atomic variables are sequentially consistent: there is a global total order among them.

A crucial goal in the design of any practical memory model is to preserve, as much as possible, the freedom of compiler writers to employ code improvement techniques originally developed for sequential programs. The ordering constraints imposed by synchronization operations necessitate not only hardware-level ordered accesses or memory fences, but also software-level "compiler fences," which inhibit the sorts of code motion traditionally used for latency tolerance, redundancy elimination, etc. (Recall that in our pseudocode synchronizing instructions are intended to enforce both hardware and compiler ordering.) Much of the complexity of C/C++ atomic variables stems from the desire to avoid unnecessary hardware ordering and compiler fences, across a variety of hardware platforms. Within reason, programmers should attempt in C/C++ to specify the minimal ordering constraints required for correct behavior. At the same time, they should resist the temptation to "get by" with minimal ordering in the absence of a solid correctness argument. In particular, while the language allows the programmer to relax the default sequential consistency of accesses to atomic variables (presumably to avoid paying for write atomicity), the result can be very confusing. Recent work by Attiya et al. [2011] has also shown that certain W‖R orderings and fetch_and_Φ operations are essential in a fundamental way: standard concurrent objects cannot be written without them.

CHAPTER 4

Practical Spin Locks

The mutual exclusion problem was first identified in the early 1960s. Dijkstra attributes the first 2-thread solution to Theodorus Dekker [Dijkstra, 1968b]. Dijkstra himself published an n-thread solution in 1965 [CACM]. The problem has been intensely studied ever since. Taubenfeld [2008] provides a summary of significant historical highlights. Ben-Ari [2006, Chaps. 3 & 5] presents a bit more detail. Much more extensive coverage can be found in Taubenfeld's encyclopedic text [Taubenfeld, 2006].

Through the 1960s and '70s, attention focused mainly on algorithms in which the only atomic primitives were assumed to be load and store. Since the 1980s, practical algorithms have all assumed the availability of more powerful atomic primitives, though interest in load/store-only algorithms continues in the theory community.

We present a few of the most important load-store-only spin locks in the first subsection below. In Section 4.2 we consider simple locks based on test_and_set (TAS) and fetch_and_increment (FAI). In Section 4.3 we turn to queue-based locks, which scale significantly better on large machines. In Section 4.4 we consider extensions of the basic acquire–release API. Finally, in Section 4.5, we consider additional techniques to reduce unnecessary overhead.

4.1 CLASSICAL LOAD-STORE-ONLY ALGORITHMS

Peterson's Algorithm

The simplest known 2-thread spin lock (Figure 4.1) is due to Peterson [1981]. The lock is represented by a pair of Boolean variables, interested[self] and interested[other] (initially false), and a integer turn that is either 0 or 1. To acquire the lock, thread i indicates its interest by setting interested[self] and then waiting until either (a) the other thread is not interested or (b) turn is set to the other thread, indicating that thread i set it first.

To release the lock, thread i sets interested[self] back to false. This allows the other thread, if it is waiting, to enter the critical section. The initial value of turn in each round is immaterial: it serves only to break the tie when both threads are interested in entering the critical section.

In his original paper, Peterson showed how to extend the lock to n threads by proceeding through a series of $n - 1$ rounds, each of which eliminates a possible contender. Total (remote-access) time for a thread to enter the critical section, however, is $\Omega(n^2)$, even in the absence of contention. In separate work, Peterson and Fischer [1977] showed how to generalize any 2-thread solution to n threads with a hierarchical *tournament* that requires only $O(\log n)$ time, even in the

```
class lock
    (0, 1) turn
    bool interested[0..1] := { false, false }
```

```
lock.acquire():                                              lock.release():
    other := 1 − self                                            interested[self].store(false, RW‖)
    interested[self].store(true)
    turn.store(self)
    while interested[other].load() and turn.load() ≠ other;    // spin
    fence(R‖RW)
```

Figure 4.1: Peterson's 2-thread spin lock. Variable self must be either 0 or 1.

presence of contention. Burns and Lynch [1980] proved that any deadlock-free mutual exclusion algorithm using only reads and writes requires $\Omega(n)$ space.

Lamport's Bakery Algorithm

Most of the n-thread mutual exclusion algorithms based on loads and stores can be shown to be starvation free. Given differences in the relative rates of progress of different threads, however, most allow a thread to be bypassed many times before finally entering the critical section. In an attempt to improve fault tolerance, Lamport [1974] proposed a "bakery" algorithm (Figure 4.2) inspired by the "please take a ticket" and "now serving" signs seen at bakeries and other service

---

Synchronizing Instructions in Peterson's Algorithm

Our code for Peterson's algorithm employs several explicit synchronizing instructions. (For a reminder of our notation, see Table 2.1, page 19.) The ‖RW fence at the end of acquire is an "acquire fence": it ensures that the lock is held (the preceding spin has completed successfully) before the thread can execute any instructions in the critical section (the code that will follow the return). Similarly, the RW‖ store in release is a "release access": it ensures that all instructions in the critical section (the code that preceded the call) have completed before the lock is released.

The two synchronizing stores in acquire are needed to ensure the correctness of the lock, which depends critically on the order in which these accesses occur. The write to interested[self] must precede the write to turn. The latter write must then precede the reads of interested[other] and turn. Without the explicit synchronizing accesses, the compiler might reorder loads and stores in the assembly language program, or the hardware might allow their effects to occur out of order from the other thread's perspective. As just one example of the trouble that might ensue, suppose that threads 0 and 1 execute acquire more or less at the same time, and that thread 0 sets turn first, but thread 0's write to interested[0] is slow to propagate to thread 1. Then thread 0 may read interested[1] = true but turn = 1, allowing it to enter the critical section, while thread 1 reads turn = 0 but interested[0] = false, allowing it to enter also.

Though we will not usually present them explicitly, similar arguments for synchronizing instructions apply to algorithms throughout the remainder of this lecture. Most lock acquire operations, for example, will end with a fence instruction that orders a prior spin with respect to the ordinary loads and stores of the upcoming critical section. Most lock release operations will begin with a synchronizing store that is ordered with respect to prior ordinary loads and stores.

```
class lock
    bool choosing[𝒯] := { false ... }
    int number[𝒯] := { 0 ... }
lock.acquire():
    choosing[self].store(true, ‖R)
    int m := 1 + max_{i∈𝒯} (L→number[i])              lock.release():
    number[self].store(m, R‖)                              number[self].store(0, RW‖)
    choosing[self].store(false)
    for i ∈ 𝒯
        while choosing[i].load();              // spin
        repeat
            int t := number[i].load()          // spin
        until t = 0 or ⟨t, i⟩ ≥ ⟨m, self⟩
    fence(R‖RW)
```

Figure 4.2: Lamport's bakery algorithm. The max operation is not assumed to be atomic. It *is*, however, assumed to read each number field only once.

counters. His algorithm has the arguably more significant advantage that threads acquire the lock in the order in which they first indicate their interest—i.e., in FIFO order.

Each thread begins by scanning the number array to find the largest "ticket" value held by a waiting thread. During the scan it sets its choosing flag to true to let its peers know that its state is in flux. After choosing a ticket higher than any it has seen, it scans the array again, spinning until each peer's ticket is (a) stable and (b) greater than or equal to its own. The second

---

Defining Time Complexity for Spin Locks

Given that we generally have no bounds on either the length of a critical section or the relative rates of execution of different threads, we cannot in general bound the number of load instructions that a thread may execute in the acquire method of any spin lock algorithm. How then can we compare the time complexity of different locks?

The standard answer is to count only accesses to shared variables (not those that are thread-local), and then only when the access is "remote." This is not a perfect measure, since local accesses are not free, but it captures the dramatic difference in cost between cache hits and misses on modern machines.

On an NRC-NUMA machine (page 13), the definition of "remote" is straightforward: we associate a (static) location with each variable and thread, and charge for all and only those accesses made by threads to data at other locations. On a globally cache-coherent machine, the definition is less clear, since whether an access hits in the cache may depend on whether there has been a recent access by another thread. The standard convention is to count all and only those accesses to shared variables that *might* be conflict misses. In a simple loop that spins on a Boolean variable, for example, we would count the initial load that starts the spin and the final load that ends it. Ideally, we would not count any of the loads in-between. If, however, another thread could write the variable and then restore its value before we read it again, we would need to consider—and count—the number of times this could happen.

Unless otherwise noted, we will use the globally cache coherent model in this lecture.

of these spins uses lexicographic comparison of ⟨value, thread id⟩ pairs to resolve any ties in the chosen tickets. The equals case in the comparison avoids the need for special-case code when a thread examines its own number field. Once its own ⟨value, thread id⟩ pair is the smallest in the array, a thread knows that it is "first in line," and can enter the critical section.

Like all deadlock-free mutual exclusion algorithms based only on loads and stores, the bakery algorithm requires $\Omega(n)$ space, where n is the total number of threads in the system. Total time to enter the critical section is also $\Omega(n)$, even in the absence of contention. As originally formulated (and as shown in Figure 4.2), number fields in the bakery algorithm grow without bound. Taubenfeld [2004] has shown how to bound them instead. For machines with more powerful atomic primitives, the conceptually similar *ticket lock* [Fischer et al., 1979, Reed and Kanodia, 1979] (Section 4.2.2) uses fetch_and_increment on shared "next ticket" and "now serving" variables to reduce space requirements to $O(1)$ and time to $O(m)$, where m is the number of threads concurrently competing for access.

Lamport's Fast Algorithm

One of the key truisms of parallel computing is that if a lock is highly contended most of the time, then the program in which it is embedded probably won't scale. Turned around, this observation suggests that in a well designed program, the typical spin lock will usually be free when a thread attempts to acquire it. Lamport's "fast" algorithm [1987] (Figure 4.3) exploits this observation by arranging for a lock to be acquired in constant time in the absence of contention (but in $O(n)$ time, where n is the total number of threads in the system, whenever contention is encountered).

The core of the algorithm is a pair of lock fields, x and y. To acquire the lock, thread t must write its id into x and then y, and be sure that no other thread has written to x in-between. Thread t checks y immediately after writing x, and checks x immediately after writing y. If y is not ⊥ when checked, some other thread must be in the critical section; t waits for it to finish and retries. If x is not still t when checked, some other thread may have entered the critical section (t cannot be sure); in this case, t must wait until the competing thread(s) have either noticed the conflict or left the critical section.

To implement its "notice the conflict" mechanism, the fast lock employs an array of trying flags, one per thread. Each thread sets its flag while competing for the lock (it also leaves it set while executing a critical section for which it encountered no contention). If a thread t *does* detect contention, it unsets its trying flag, waits until the entire array is clear, and then checks to see if it was the *last* thread to set y. If so, it enters the critical section (after first executing a ∥RW fence to order upcoming ordinary accesses with respect to previous synchronizing instructions). If not, it retries the acquisition protocol.

A disadvantage of the fast lock as presented (both here and in the original paper) is that it requires $\Omega(n)$ time (where n is the total number of threads in the system) even when only two threads are competing for the lock. Building on the work of Moir and Anderson [1995], Merritt

```
class lock
    T x
    T y := ⊥
    bool trying[T] := { false ... }
lock.acquire():                                                    lock.release():
    loop                                                               y.store(⊥, RW‖)
        trying[self].store(true)                                       trying[self].store(false)
        x.store(self)
        if y.load() ≠ ⊥
            trying[self].store(false)
            while y.load() ≠ ⊥;          // spin
            continue                     // go back to top of loop
        y.store(self)
        if x.load() ≠ self
            trying[self].store(false)
            for i ∈ T
                while trying[i].load();  // spin
            if y.load() ≠ self
                while y.load() ≠ ⊥;      // spin
                continue                 // go back to top of loop
        break
    fence(R‖RW)
```

Figure 4.3: Lamport's fast algorithm.

and Taubenfeld [2000] show how to reduce this time to $O(m)$, where m is the number of threads concurrently competing for access.

4.2   CENTRALIZED ALGORITHMS

As noted in Sections 1.3 and 2.3, almost every modern machine provides read-modify-write (fetch_and_Φ) instructions that can be used to implement mutual exclusion in constant space and—in the absence of contention—constant time. The locks we will consider in this section—all of which use such instructions—are *centralized* in the sense that they spin on a single, central location. They differ in fairness and in the performance they provide in the presence of contention.

4.2.1   TEST_AND_SET LOCKS

The simplest mutex spin lock embeds a TAS instruction in a loop, as shown in Figure 4.4. On a typical machine, the TAS instruction will require a writable copy of the target location, necessitating communication across the processor-memory interconnect on every iteration of the loop. These messages will typically serialize, inducing enormous hardware contention that interferes

```
class lock
    bool f := false

lock.acquire():                        lock.release():
    while TAS(&f);    // spin              f.store(false, RW‖)
    fence(R‖RW)
```

Figure 4.4: The simple test_and_set lock.

```
class lock
    bool f := false

lock.acquire():                        lock.release():
    while TAS(&f)                          f.store(false, RW‖)
        while f;      // spin
    fence(R‖RW)
```

Figure 4.5: The test-and-test_and_set lock. Unlike the test_and_set lock of Figure 4.4, this code will typically induce interconnect traffic only when the lock is modified by another core.

not only with other threads that are attempting to acquire the lock, but also with any attempt by the lock owner to release the lock.

Performance can be improved by arranging to obtain write permission on the lock only when it appears to be free. Proposed by Rudolph and Segall [1984], this *test-and*-test_and_set lock is still extremely simple (Figure 4.5), and tends to perform well on machines with a small handful of cores. Whenever the lock is released, however, every competing thread will fall out of its inner loop and attempt another TAS, each of which induces coherence traffic. With n threads continually attempting to execute a critical sections, total time *per* acquire-release *pair* will be $O(n)$, which is still unacceptable on a machine with more than a handful of cores.

Drawing inspiration from the classic Ethernet contention protocol [Metcalfe and Boggs, 1976], Anderson et al. [1990] proposed an exponential backoff strategy for test_and_set locks (Figure 4.6). Experiments indicate that it works quite well in practice, leading to near-constant overhead per acquire-release pair on many machines. Unfortunately, it depends on constants (the base, multiplier, and limit for backoff) that have no single best value in all situations. Ideally, they should be chosen individually for each machine and workload. Note that test_and_set suffices in the presence of backoff; test-and-test_and_set is not required.

4.2.2 THE TICKET LOCK

Test_and_set locks are potentially unfair. While most machines can be expected to "randomize" the behavior of TAS (e.g., so that some particular core doesn't always win when more than one attempts a TAS at roughly the same time), and while exponential backoff can be expected to inject additional variability into the lock's behavior, it is still entirely possible for a thread that has

```
class lock
    bool f := false
    const int base = ...              // tuning parameters
    const int limit = ...
    const int multiplier = ...
lock.acquire():
    int delay := base                                lock.release():
    while TAS(&f)                                         f.store(false, RW‖)
        pause(delay)
        delay := min(delay × multiplier, limit)
    fence(R‖RW)
```

Figure 4.6: The test_and_set lock with exponential backoff. The pause(k) operation is typically an empty loop that iterates k times. Ideal choices of base, limit and multiplier values depend on the machine architecture and, typically, the application workload.

```
class lock
    int next_ticket := 0
    int now_serving := 0
    const int base = ...              // tuning parameter
lock.acquire():
    int my_ticket := FAI(&next_ticket)
        // returns old value;
        // arithmetic overflow is harmless               lock.release():
    loop                                                     int t := now_serving + 1
        int ns := now_serving.load()                         now_serving.store(t, RW‖)
        if ns = my_ticket
            break
        pause(base × (my_ticket − ns))
            // overflow in subtraction is harmless
    fence(R‖RW)
```

Figure 4.7: The ticket lock with proportional backoff. Tuning parameter base should be chosen to be roughly the length of a trivial critical section.

been waiting a very long time to be passed up by a relative newcomer; in principle, a thread can starve.

The *ticket lock* [Fischer et al., 1979, Reed and Kanodia, 1979] (Figure 4.7) addresses this problem. Like Lamport's bakery lock, it grants the lock to competing threads in first-come-first-served order. Unlike the bakery lock, it uses fetch_and_increment to get by with constant space, and with time (per lock acquisition) roughly linear in the number of competing threads.

The code in Figure 4.7 employs a backoff strategy due to Mellor-Crummey and Scott [1991b]. It leverages the fact that my_ticket − L.now_serving represents the number of threads ahead of the calling thread in line. If those threads consume an average of $k \times$ base time per critical section, the calling thread can be expected to probe now_serving about k times before acquiring the lock. Under high contention, this can be substantially smaller than the $O(n)$ probes expected without backoff.

In a system that runs long enough, the next_ticket and now_serving counters can be expected to exceed the capacity of a fixed word size. Rollover is harmless, however: the maximum number of threads in any reasonable system will be less than the largest representable integer, and subtraction works correctly in the ring of integers mod $2^{wordsize}$.

4.3   QUEUED SPIN LOCKS

Even with proportional backoff, a thread can perform an arbitrary number of remote accesses in the process of acquiring a ticket lock, inducing an arbitrary amount of contention. Anderson et al. [1990] and (independently) Graunke and Thakkar [1990] showed how to reduce this to a small constant on a globally cache-coherent machine. The intuition is to replace a single now_serving variable (or the Boolean flag of a test_and_set lock) with a queue of waiting threads. Each thread knows its place in line: it waits for its predecessor to finish before entering the critical section, and signals its successor when it's done.

Both Anderson et al. and Graunke and Thakkar implement the queue as an array of n flag words, where n is the maximum number of threads in the system. Both arrange for every thread to spin on a different element of the array, and to know the index of the element on which its successor is spinning. In Anderson et al.'s lock, array elements are allocated dynamically using fetch_and_increment; a thread releases the lock by updating the next element (in circular order) after the one on which it originally spun. In Graunke and Thakkar's lock, array elements are statically allocated. A thread releases the lock by writing its own element; it finds the element on which to spin by performing a swap on an extra tail element.

Inspired by the QOLB hardware primitive of the Wisconsin Multicube [Goodman et al., 1989] and the IEEE SCI standard [Aboulenein et al., 1994] (Section 2.3.2), Mellor-Crummey and Scott [1991b] devised a queue-based spin lock that employs a linked list instead of an array. Craig and, independently, Magnussen, Landin, and Hagersten devised an alternative version that, in essence, links the list in the opposite direction. Unlike the locks of Anderson et al. and Graunke and Thakkar, these list-based locks do not require a static bound on the maximum number of threads; equally important, they require only $O(n + j)$ space for n threads and j locks, rather than $O(nj)$. They are generally considered the methods of choice for FIFO locks on large-scale systems. (Note, however, that strict FIFO ordering may be inadvisable on a system with preemption; see Section 7.5.2.)

4.3.1   THE MCS LOCK

Pseudocode for the MCS lock appears in Figure 4.8. Every thread using the lock allocates a qnode record containing a queue link and a Boolean flag. Typically, this record lies in the stack frame of the code that calls acquire and release; it must be passed as an argument to both (but see the discussion under "Modifications for a Standard Interface" below).

```
type qnode = record
    qnode* next
    bool waiting
class lock
    qnode* tail := null
```

```
lock.acquire(qnode* p):                                    // Initialization of waiting can be delayed
    p→next := null                                         // until the if statement below,
    p→waiting := true                                      // but at the cost of an extra W‖W fence.
    qnode* prev := swap(&tail, p, W‖)
    if prev ≠ null                                         // queue was nonempty
        prev→next.store(p)
        while p→waiting.load();                            // spin
    fence(R‖RW)
```

```
lock.release(qnode* p):
    qnode* succ := p→next.load(WR‖)
    if succ = null                                         // no known successor
        if CAS(&tail, p, null) return
        repeat succ := p→next.load() until succ ≠ null
    succ→waiting.store(false)
```

Figure 4.8: The MCS queued lock.

Threads holding or waiting for the lock are chained together, with the link in the qnode of thread *t* pointing to the qnode of the thread to which *t* should pass the lock when done with its critical section. The lock itself is simply a pointer to the qnode of the thread at the tail of the queue, or null if the lock is free.

Operation of the lock is illustrated in Figure 4.9. The acquire method allocates a new qnode, initializes its next pointer to null, and swaps it into the tail of the queue. If the value returned by the swap is null, then the calling thread has acquired the lock (line 2). If the value returned by the swap is non-null, it refers to the qnode of the caller's predecessor in the queue (indicated by the dashed arrow in line 3). Here thread B must set A's next pointer to refer to its own qnode. Meanwhile, some other thread C may join the queue (line 4).

When thread A has completed its critical section, the release method reads the next pointer of A's qnode to find the qnode of its successor B. It changes B's waiting flag to false, thereby granting it the lock (line 5).

If release finds that the next pointer of its qnode is null, it attempts to CAS the lock tail pointer back to null. If some other thread has already swapped itself into the queue (line 5), the CAS will fail, and release will wait for the next pointer to become non-null (line 6). If there are no waiting threads (line 7), the CAS will succeed, returning the lock to the appearance in line 1.

The MCS lock has several important properties. Threads join the queue in a wait-free manner (using swap), after which they receive the lock in FIFO order. Each waiting thread spins on a separate location, eliminating contention for cache and interconnect resources. In fact, because each thread allocates its own qnode, it can arrange for it to be local even on an NRC-

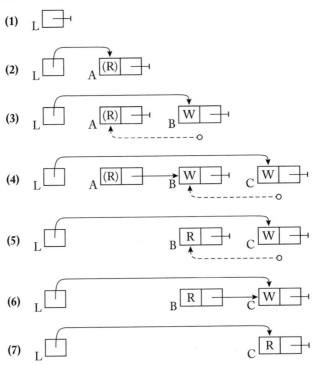

Figure 4.9: Operation of the MCS lock. An 'R' indicates that the thread owning the given qnode is running its critical section (parentheses indicate that the value of the waiting flag is immaterial). A 'W' indicates that the corresponding thread is waiting. A dashed arrow represents a local pointer (returned to the thread by swap).

NUMA machine. Total (remote access) time to pass the lock from one thread to the next is constant. Total space is linear in the number of threads and locks.

As written (Figure 4.8), the MCS lock requires both swap and CAS. CAS can of course be used to emulate the swap in the acquire method, but entry to the queue drops from wait-free to lock-free (meaning that a thread can theoretically starve). Mellor-Crummey and Scott [1991b] also show how to make do with only swap in the release method, but FIFO ordering may be lost when a thread enters the queue just as its predecessor is releasing the lock.

Modifications for a Standard Interface
One disadvantage of the MCS lock is the need to pass a qnode pointer to acquire and release. Test_and_set and ticket locks pass only a reference to the lock itself. If a programmer wishes to convert code from traditional to queued locks, or to design code in which the lock implementation can be changed at system configuration time, it is natural to wish for a version of the MCS

lock that omits the extra parameters, and can be substituted in without rewriting all the call points. Auslander et al. [2003] devised such a version as part of the K42 project at IBM Research [Appavoo et al., 2005]. Their code exploits the fact that once a thread has acquired a lock, its qnode serves only to hold a reference to the next thread in line. Since the thread now "owns" the lock, it can move its next pointer to an extra field of the lock, at which point the qnode can be discarded.

Code for the K42 variant of the MCS lock appears in Figure 4.10. Operation of the lock is illustrated in Figure 4.11. An idle, unheld lock is represented by a qnode containing two null pointers (line 1 of Figure 4.11). The first of these is the usual tail pointer from the MCS lock; the other is a "next" pointer that will refer to the qnode of the first waiting thread, if and when there is one. Newly arriving thread A (line 2) uses CAS to replace a null tail pointer with a pointer to the lock variable itself, indicating that the lock is held, but that no other threads are waiting. At this point, newly arriving thread B (line 3) will see the lock variable as its "predecessor," and will update the next field of the lock rather than that of A's qnode (as it would have in a regular MCS lock). When thread C arrives (line 4), it updates B's next pointer, because it obtained a pointer to B's qnode when it performed a CAS on the tail field of the lock. When A completes its critical section (line 5), it finds B's qnode by reading the head field of the lock. It then changes B's "head" pointer (which serves as a waiting flag) to null, thereby releasing B. Upon leaving its spin, B updates the head field of the lock to refer to C's qnode. Assuming no other threads arrive, when C completes its critical section it will return the lock to the state shown in line 1.

The careful reader may notice that the code of Figure 4.10 has a lock-free (not wait-free) entry protocol, and thus admits the (remote, theoretical) possibility of starvation. This can be remedied by replacing the original CAS with a swap, but a thread that finds that the lock was previously free must immediately follow up with a CAS, leading to significantly poorer performance in the (presumably common) uncontended case. A potentially attractive hybrid strategy starts with a load of the tail pointer, following up with a CAS if the lock appears to be free and a swap otherwise.

4.3.2 THE CLH LOCK

Because every thread spins on a field of its own qnode, the MCS lock achieves a constant bound on the number of remote accesses per lock acquisition, even on a NRC-NUMA machine. The cost of this feature is the need for a newly arriving thread to write the address of its qnode into the qnode of its predecessor, and for the predecessor to wait for that write to complete before it can release a lock whose tail pointer no longer refers to its own qnode.

Craig [1993] and, independently, Magnussen, Landin, and Hagersten [1994] observed that this extra "handshake" can be avoided by arranging for each thread to spin on its predecessor's qnode, rather than its own. On a globally cache-coherent machine, the spin will still be local, because the predecessor's node will migrate to the successor's cache. The downside of the change is that a thread's qnode must potentially remain accessible long after the thread has left its critical

```
type qnode = record
    qnode* tail
    qnode* next
const qnode* waiting = 1
    // In a real qnode, tail = null means the lock is free;
    // In the qnode that is a lock, tail is the real tail pointer.
class lock
    qnode q := { null, null }

lock.acquire():
    loop
        qnode* prev := q.tail.load()
        if prev = null                              // lock appears to be free
            if CAS(&q.tail, null, &q) break
        else
            qnode n := { waiting, null }
            if CAS(&q.tail, prev, &n, W∥)            // we're in line
                prev→next.store(&n)
                while n.tail.load() = waiting;       // spin
                // now we have the lock
                qnode* succ := n.next.load()
                if succ = null
                    q.next.store(null)
                    // try to make lock point at itself:
                    if ¬CAS(&q.tail, &n, &q)
                        // somebody got into the timing window
                        repeat succ := n.next.load() until succ ≠ null
                        q.next.store(succ)
                    break
                else
                    q.next.store(succ)
                    break
    fence(R∥RW)

lock.release():
    qnode* succ := q.next.load(RW∥)
    if succ = null
        if CAS(&q.tail, &q, null) return
        repeat succ := q.next.load() until succ ≠ null
    succ→tail.store(null)
```

Figure 4.10: K42 variant of the MCS queued lock. Note the standard interface to acquire and release, with no parameters other than the lock itself.

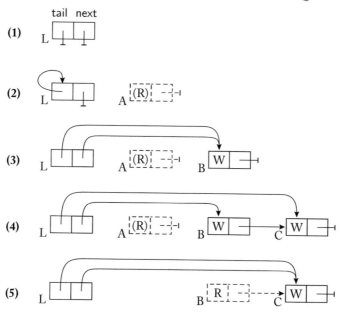

Figure 4.11: Operation of the K42 MCS lock. An 'R' (running) indicates a null "tail" pointer; 'W' indicates a "waiting" flag. Dashed boxes indicate qnodes that are no longer needed, and may safely be freed by returning from the method in which they were declared. In (1) the lock is free. In (2) a single thread is active in the critical section. In (3) and (4) two new threads have arrived. In (5) thread A has finished and thread B is now active.

section: we cannot bound the time that may elapse before a successor needs to inspect that node. This requirement is accommodated by having a thread provide a fresh qnode to acquire, and return with a *different* qnode from release.

In their original paper, Magnussen, Landin, and Hagersten presented two versions of their lock: a simpler "LH" lock and an enhanced "M" lock; the latter reduces the number of cache misses in the uncontended case by allowing a thread to keep its original qnode when no other thread is trying to acquire the lock. The M lock needs CAS to resolve the race between a thread that is trying to release a heretofore uncontended lock and the arrival of a new contender. The LH lock has no such races; all it needs is swap.

Craig's lock is essentially identical to the LH lock: it differs only in the mechanism used to pass qnodes to and from the acquire and release methods. It has become conventional to refer to this joint invention by the initials of all three inventors: CLH.

Code for the CLH lock appears in Figure 4.12. An illustration of its operation appears in Figure 4.13. A free lock (line 1 of the latter figure) contains a pointer to a qnode whose succ_must_wait flag is false. Newly arriving thread A (line 2) obtains a pointer to this node

```
type qnode = record
      qnode* prev                              // read and written only by owner thread
      bool succ_must_wait
class lock
      qnode dummy := { null, false }
      // ideally, dummy and tail should lie in separate cache lines
      qnode* tail := &dummy

lock.acquire(qnode* p):
      p→succ_must_wait := true
      qnode* pred := p→prev := swap(&tail, p, W‖)
      while pred→succ_must_wait.load();                    // spin
      fence(R‖RW)

lock.release(qnode** pp):
      qnode* pred := (*pp)→prev
      (*pp)→succ_must_wait.store(false, RW‖)
      *pp := pred                                          // take pred's qnode
```

Figure 4.12: The CLH queued lock.

(dashed arrow) by executing a swap on the lock tail pointer. It then spins on this node (or simply observes that its succ_must_wait flag is already false). Before returning from acquire, it stores the pointer into its own qnode so it can find it again in release. (In the LH version of the lock [Magnussen, Landin, and Hagersten, 1994], there was no pointer in the qnode; rather, the API for acquire returned a pointer to the predecessor qnode as an explicit parameter.)

To release the lock (line 4), thread A writes false to the succ_must_wait field of its own qnode and then leaves that qnode behind, returning with its predecessor's qnode instead (here previously marked 'X'). Thread B, which arrived at line 3, releases the lock in the same way. If no other thread is waiting at this point, the lock returns to the state in line 1.

In his original paper, Craig [1993] explored several extensions to the CLH lock. By introducing an extra level of indirection, one can eliminate remote spinning even on an NRC-NUMA machine—without requiring CAS, and without abandoning strict either FIFO ordering or wait-free entry. By linking the list both forward and backward, and traversing it at acquire time, one can arrange to grant the lock in order of some external notion of priority, rather than first-come-first-served (Markatos [1991] presented a similar technique for MCS locks). By marking nodes as abandoned, and skipping over them at release time, one can accommodate timeout (we will consider this topic further in Section 7.5.2, together with the possibility—suggested by Craig as future work—of skipping over threads that are currently preempted). Finally, Craig sketched a technique to accommodate nested critical sections without requiring a thread to allocate multiple qnodes: arrange for the thread to acquire its predecessor's qnode when the lock is acquired rather than when it is released, and maintain a separate thread-local stack of pointers to the qnodes that must be modified in order to release the locks.

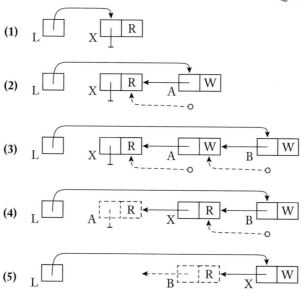

Figure 4.13: Operation of the CLH lock. An 'R' indicates that a thread spinning on this qnode (i.e., the successor of the thread that provided it) is free to run its critical section; a 'W' indicates that it must wait. Dashed boxes indicate qnodes that are no longer needed by successors, and may be reused by the thread releasing the lock. Note the change in label on such nodes, indicating that they now "belong" to a different thread.

Modifications for a Standard Interface

Craig's suggestion for nested critical sections requires that locks be released in the reverse of the order in which they were acquired; it does not generalize easily to idioms like hand-over-hand locking (Section 3.1.2). If we adopt the idea of a head pointer field from the K42 MCS lock, however, we can devise a (previously unpublished) CLH variant that serves as a "plug-in" replacement for traditional locks (Figure 4.14).

Our code assumes a global array, thread_qnode_ptrs, indexed by thread id. In practice this could be replaced by any form of "thread-local" storage—e.g., the Posix pthread_getspecific mechanism. Operation is very similar to that of the original CLH lock: the only real difference is that instead of requiring the caller to pass a qnode to release, we leave that pointer in a head field of the lock. No dynamic allocation of qnodes is required: the total number of extant nodes is always $n + j$ for n threads and j locks—one per lock at the head of each queue (with succ_-must_wait true or false, as appropriate), one enqueued (not at the head) by each thread currently waiting for a lock, and one (in the appropriate slot of thread_qnode_ptrs) reserved for future use by each thread not currently waiting for a lock. (Elements of thread_qnode_ptrs corresponding

```
qnode initial_thread_qnodes[𝒯]
qnode* thread_qnode_ptrs[𝒯] := { i ∈ 𝒯: &initial_thread_qnodes[i] }
type qnode = record
    bool succ_must_wait
class lock
    qnode dummy := { false }
    // ideally, dummy should lie in a separate cache line from tail and head
    qnode* tail := &dummy
    qnode* head

lock.acquire():
    qnode* p := thread_qnode_ptrs[self]
    p→succ_must_wait := true
    qnode* pred := swap(&tail, p, W‖)
    while pred→succ_must_wait.load();          // spin
    head.store(p)
    thread_qnode_ptrs[self] := pred
    fence(R‖RW)

lock.release():
    head→succ_must_wait.store(false, RW‖)
```

Figure 4.14: A CLH variant with standard interface.

to threads currently waiting for a lock are overwritten, at the end of acquire, before being read again.)

4.3.3   WHICH SPIN LOCK SHOULD I USE?

On modern machines, there is little reason to consider load-store-only spin locks, except perhaps as an optimization in programs with highly asymmetric access patterns (see Section 4.5.3 below).

Table 4.1 summarizes the tradeoffs among fetch_and_Φ-based spin locks. Absolute performance will vary with the machine and workload, and is difficult to quantify here. For small numbers of threads—single digits, say—both the test_and_set lock with exponential backoff and the ticket lock with proportional backoff tend to work quite well. The ticket lock is fairer, but this can actually be a disadvantage in some situations (see the discussion of locality-conscious locking in Section 4.5.1 and of inopportune preemption in Section 7.5.2).

The problem with both test_and_set and ticket locks is their brittle performance as the number of contending threads increases. In any application in which lock contention may be a bottleneck—even rarely—it makes sense to use a queue-based lock. Here the choice between MCS and CLH locks depends on architectural features and costs. The MCS lock is generally preferred on an NRC-NUMA machine: the CLH lock can be modified to avoid remote spinning, but the extra level of indirection requires additional fetch_and_Φ operations on each lock transfer. For machines with global cache coherence, either lock can be expected to work well. Given the

Table 4.1: Tradeoffs among fetch_and_Φ-based spin locks. Symbols are meant to suggest "good" (+), "fair" (○), and "poor" (−). Space needs are in words, for n threads and j locks, none of which is requested or held.

	TAS	ticket (w/ backoff)	MCS	CLH (original)	MCS	CLH ("K42")
fairness	−	+	+	+	+	+
preemption tolerance	○	−	−	−	−	−
scalability	○	○	+	+	+	+
fast-path overhead	+	+	○	○	−	−
interoperability	+	+	−	−	+	+
NRC-NUMA suitability	−	−	+	○	+	○
space needs	j	$2j$	j	$2j$	$2j$	$2n + 3j$

absence of dummy nodes, space needs are lower for MCS locks, but performance may be better (by a small constant factor) for CLH locks on some machines.

4.4   INTERFACE EXTENSIONS

The acquire–release interface for locks—both busy-wait and blocking—is often extended to accommodate special use cases. In many packages, for example, acquire takes an extra, optional *timeout* parameter that specifies the maximum length of time that the thread is willing to wait to acquire the lock. A Boolean return value then indicates whether the lock was actually acquired, or whether it timed out. Timeout is sometimes useful in interactive applications, which may need to pursue an alternative code path when a lock is not available in time to meet a real-time deadline. It may also be useful in programs that are prone to deadlock, and in which prevention and avoidance (Section 3.1.1) are impractical: expiration of a lengthy timeout can be taken as an indication that deadlock has probably occurred, and that the calling thread should back out, release its other locks, and retry the operation later.

In some cases, an application may wish to specify a "timeout" of zero—to acquire a lock if and only if it is not currently held. This *trylock* extension is straightforward: a try-acquire operation takes no extra parameters, but does return a Boolean: true indicates that the lock has just been acquired; false indicates that it was already held by another thread.

Throughout this chapter we have been assuming that the access pattern for a given lock is always a sequence of acquire-release pairs, in which the release method is called by the same thread that called acquire—and before the thread attempts to acquire the same lock again. This assumption is not explicit in the acquire–release API, but is often embedded in the underlying implementation.

There are times, however, when it may be desirable to allow a thread to acquire the same lock multiple times, so long as it releases it the same number of times before any other thread acquires it. Suppose, for example, that operation foo accesses data protected by lock L, and that foo is sometimes called by a thread that already holds L, and sometimes by a thread that does not. In the latter case, foo needs to acquire L. With most of the locks presented above, the program will deadlock in the former case: the thread will end up "waiting for itself."

A lock that can be re-acquired by the owner thread, and that returns to being free only after an equal number of releases, is said to be *reentrant*. A simple strategy, which can be used with any mutual exclusion lock (spin or scheduler-based), is to extend the base implementation with an owner field and a counter:

```
class reentrant_lock
    lock L
    int owner := none
    int count := 0

reentrant_lock.acquire()              reentrant_lock.release()
    if owner ≠ self                       if −−count = 0
        L.acquire()                           owner := none
        owner := self                         L.release()
    count++
```

Given the overhead of inspecting and updating owner and count fields, many designers choose not to make locks reentrant by default.

The astute reader may notice that the read of owner in reentrant_lock.acquire races with the writes of owner in both acquire and release. In memory models that forbid data races, the owner field may need to be declared as volatile or atomic.

4.5   SPECIAL-CASE OPTIMIZATIONS

Many techniques have been proposed to improve the performance of spin locks in important special cases. We will consider the most important of these—read-mostly synchronization—in Chapter 6. In this section we consider three others. Locality-conscious locking biases the acquisition of a lock toward threads that are physically closer to the most recent prior holder, thereby reducing average hand-off cost on NUMA machines. Double-checked locking addresses situations in which initialization of a variable must be synchronized, but subsequent use need not be. Asymmetric locking addresses situations in which a lock is accessed repeatedly by the same thread, and performance may improve if that thread is able to reacquire the lock more easily than others can acquire it.

4.5.1   LOCALITY-CONSCIOUS LOCKING

On a NUMA machine—or even one with a non-uniform cache architecture (sometimes known as NUCA)—inter-core communication costs may differ dramatically. If, for example, we have multiple processors, each with multiple cores, we may be able to pass a lock to another core within the same processor much faster than we can pass it to a core of another processor. More significantly, since locks are typically used to protect shared data structures, we can expect the cache lines of the protected structure to migrate to the acquiring core, and this migration will be cheaper if the core is nearby rather than remote.

Radović and Hagersten [2002] were the first to observe the importance of locality in locking, and to suggest passing locks to nearby cores when possible. Their "RH lock," developed for a machine with two NUMA "clusters," is essentially a pair of test_and_set locks, one on each cluster, with one initialized to FREE and the other to REMOTE. A thread attempts to acquire the lock by swapping its id into its own cluster's lock. If it gets back FREE or L_FREE (locally free), it has succeeded. If it gets back a thread id, it backs off and tries again. If it gets back REMOTE, it has become the local representative of its cluster, in which case it spins (with a different set of backoff parameters) on the *other* cluster's lock, attempting to CAS it from FREE to REMOTE. To release the lock, a thread usually attempts to CAS it from its own id to FREE. If this fails, a nearby thread must be spinning, in which case the releasing thread stores L_FREE to the lock. Occasionally (subject to a tuning parameter), a releasing thread immediately writes FREE to the lock, allowing it to be grabbed by a remote contender, even if there are nearby ones as well.

While the RH lock could easily be adapted to larger numbers of clusters, space consumption would be linear in the number of such clusters—a property Radović and Hagersten considered undesirable. Their subsequent "hierarchical backoff" (HBO) [Radović and Hagersten, 2003] lock relies on statistics instead. In effect, they implement a test_and_set lock with CAS, in such a way that the lock variable indicates the cluster in which the lock currently resides. Nearby and remote threads then use different backoff parameters, so that nearby threads are more likely than remote threads to acquire the lock when it is released.

While a test_and_set lock is naturally unfair (and subject to the theoretical possibility of starvation), the RH and HBO locks are likely to be even less fair in practice. Ideally, one would like to be able to explicitly balance fairness against locality. Toward that end, Dice et al. [2012] present a general NUMA-aware design pattern that can be used with (almost) any underlying locks, including (FIFO) queued locks. Their *cohort* locking mechanism employs a global lock that indicates which cluster currently owns the lock, and a local lock for each cluster that indicates the owning thread. The global lock needs to allow release to be called by a different thread from the one that called acquire; the local lock needs to be able to tell, at release time, whether any other local thread is waiting. Apart from these requirements, cohort locking can be used with any known form of lock. Experimental results indicate particularly high throughput (and excellent fairness, subject to locality) using MCS locks at both the global and cluster level.

While the techniques discussed here improve locality only by controlling the order in which threads acquire a lock, it is also possible to control *which threads* perform the operations protected by the lock, and to assign operations that access similar data to the same thread, to minimize cache misses. Such locality-conscious allocation of work can yield major performance benefits in systems that assign fine-grain computational *tasks* to worker threads, a possibility that will arise

```
foo* p := null                          foo* p := null
lock L                                  lock L
foo* get_p():                           foo* get_p():
    L.acquire()                             foo *rtn := p.load(‖RW)
        if p = null                         if rtn = null
            p := new foo()                      L.acquire()
        foo* rtn := p                               rtn := p
    L.release()                                 // double check:
    return rtn                                  if rtn = null
                                                    rtn := new foo()
                                                    p.store(rtn, W‖)
                                            L.release()
                                        return rtn
```

Figure 4.15: Lazy initialization (left) and double-checked locking idiom (right).

in Sections 5.3.3, 7.4.3, and 8.6.3. It is also a key feature of *flat combining*, which we will consider briefly in Section 5.4.

4.5.2   DOUBLE-CHECKED LOCKING

Many applications employ shared variables that must be initialized before they are used for the first time. A canonical example can be seen on the left side of Figure 4.15. Unfortunately, this idiom imposes the overhead of acquiring L on every call to get_p. With care, we can use the *double-checked locking* idiom instead (right side of Figure 4.15).

The synchronizing accesses in the idiom are critical. In Java, variable p must be declared volatile; in C++, atomic. Without the explicit ordering, the initializing thread may set p to point to not-yet-initialized space, or a reading thread may use fields of p that were prefetched before initialization completed. On machines with highly relaxed memory models (e.g., ARM and POWER), the cost of the synchronizing accesses may be comparable to the cost of locking in the original version of the code, making the "optimization" of limited benefit. On machines with a TSO memory model (e.g., the x86 and SPARC), the optimization is much more appealing, since R‖R, R‖W, and W‖W orderings are free. Used judiciously, (e.g., in the Linux kernel for x86), double-checked locking can yield significant performance benefits. Even in the hands of experts, however, it has proven to be a significant source of bugs: with all the slightly different forms the idiom takes in different contexts, one can easily forget a necessary synchronizing access [Bacon et al., 2001]. Windows Vista introduced a special InitOnce API that regularizes usage and encapsulates all necessary ordering.

4.5.3   ASYMMETRIC LOCKING

Many applications contain data structures that are usually—or even always—accessed by a single thread, but are nonetheless protected by locks, either because they are *occasionally* accessed by another thread, or because the programmer is preserving the ability to reuse code in a future parallel context. Several groups have developed locks that can be *biased* toward a particular thread, whose acquire and release operations then proceed much faster than those of other threads. The HotSpot Java Virtual Machine, for example, uses biased locks to accommodate objects that appear to "belong" to a single thread, and to control re-entry to the JVM by threads that have escaped to native code, and may need to synchronize with a garbage collection cycle that began while they were absent [Dice et al., 2001, Russell and Detlefs, 2006].

On a sequentially consistent machine, one might be tempted to avoid (presumably expensive) fetch_and_Φ operations by using a two-thread load-store-only synchronization algorithm (e.g., Dekker's or Peterson's algorithm) to arbitrate between the preferred (bias-holding) thread and some representative of the other threads. Code might look like this:

```
class lock
    Peterson_lock PL
    general_lock GL

lock.acquire():                        lock.release():
    if ¬preferred_thread                   PL.release()
        GL.acquire()                       if ¬preferred_thread
    PL.acquire()                               GL.release()
```

The problem, of course, is that load-store-only acquire routines invariably contain some variant of the Dekker store–load sequence—

```
interested[self] := true               // store
bool potential_conflict := interested[other]   // load
if potential_conflict ...
```

—and this code works correctly on a non-sequentially consistent machine only when augmented with (presumably also expensive) W‖R ordering between the first and second lines. The cost of the ordering has led several researchers [Dice et al., 2001, Russell and Detlefs, 2006, Vasudevan et al., 2010] to propose *asymmetric* Dekker-style synchronization. Applied to Peterson's lock, the solution looks as shown in Figure 4.16.

The key is the handshake operation on the "slow" (non-preferred) path of the lock. This operation must interact with execution on the preferred thread's core in such a way that

1. if the preferred thread set fast_interested before the interaction, then the non-preferred thread is guaranteed to see it afterward.

2. if the preferred thread did *not* set fast_interested before the interaction, then it (the preferred thread) is guaranteed to see slow_interested afterward.

Handshaking can be implemented in any of several ways, including cross-core interrupts, migration to or from the preferred thread's core, forced un-mapping of pages accessed in the critical section, waiting for an interval guaranteed to contain a W‖R fence (e.g., a scheduling quantum), or explicit communication with a "helper thread" running on the preferred thread's

```
class lock
    bool fast_turn := true
    bool fast_interested := false
    bool slow_interested := false
    general_lock GL

lock.acquire():
    if preferred_thread
        fast_interested := true
        fence(W‖W)                              // free on TSO machine
        fast_turn := true
        // W‖R fence intentionally omitted
        while slow_interested and fast_turn;    // spin
    else
        GL.acquire()
        slow_interested := true
        fence(W‖W)
        fast_turn := false
        fence(W‖R)                              // slow
        handshake()                             // very slow
        while fast_interested and ¬fast_turn;   // spin
    fence(R‖RW)                                  // free on TSO machine

lock.release():
    fence(RW‖W)                                  // free on TSO machine
    if preferred_thread
        fast_interested := false
    else
        GL.release()
        slow_interested := false
```

Figure 4.16: An asymmetric lock built around Peterson's algorithm. We have written the code entirely with fences rather than annotated synchronizing accesses to highlight the ordering operations. The handshake operation on the slow path forces a known ordering with respect to the store–load sequence on the fast path.

core. Dice et al. [2001] explore many of these options in detail. Because of their cost, they are profitable only in cases where access by non-preferred threads is exceedingly rare. In subsequent work, Dice et al. [2003] observe that handshaking can be avoided if the underlying hardware provides coherence at word granularity, but supports atomic writes at subword granularity.

CHAPTER 5

Busy-wait Synchronization with Conditions

In Chapter 1 we suggested that almost all synchronization serves to achieve either atomicity or condition synchronization. Chapter 4 considered spin-based atomicity. The current chapter considers spin-based condition synchronization—flags and barriers in particular. In many cases, spins can be replaced, straightforwardly, with waits on scheduler-based synchronization queues. We will return to the topic of scheduler-based conditions in Chapter 7.

5.1 FLAGS

In its simplest form, a flag is Boolean variable, initially false, on which a thread can wait:

```
class flag
    bool f := false
flag.set():
    f.store(true, RW‖)
flag.await():
    while ¬f.load();    // spin
    fence(R‖RW)
```

Methods set and await are presumably called by different threads. Code for set consists of a release-annotated store; await ends with an acquire fence. These reflect the fact that one typically uses set to indicate that previous operations of the calling thread (e.g., initialization of a shared data structure) have completed; one typically uses await to ensure that subsequent operations of the calling thread do not begin until the condition holds.

In some algorithms, it may be helpful to have a reset method:

```
flag.reset():
    f.store(false, ‖W)
```

Before calling reset, a thread must ascertain (generally through application-specific means) that no thread is still using the flag for its previous purpose. The ‖W ordering on the store ensures that any subsequent updates (to be announced by a future set) are seen to happen after the reset.

In an obvious generalization of flags, one can arrange to wait on an arbitrary predicate:

```
class predicate
    abstract bool eval()
        // to be extended by users
predicate.await():
    while ¬eval();         // spin
    fence(R‖RW)
```

With compiler or preprocessor support, this can become

```
await( condition ):
    while ¬condition;      // spin
    fence(R‖RW)
```

This latter form is the notation we employed in Chapters 1 and 3. It must be used with care: the absence of an explicit set method means there is no obvious place to specify the release ordering that typically accompanies the setting of a Boolean flag. In any program that spins on nontrivial conditions, a thread that changes a variable that may contribute to such a condition may need to declare the variable as volatile or atomic, or update it with a RW‖ store. One must also consider the atomicity of attempts to check the condition, and the monotonicity of the condition itself: an await will generally be safe if the condition will become true due to a single store in some other thread, and never again become false. Without such a guarantee, it is unclear what can safely be assumed by the code that follows the await. We will return to generalized await statements when we consider conditional critical regions in Section 7.4.1.

5.2   BARRIER ALGORITHMS

Many applications—simulations in particular—proceed through a series of *phases*, each of which is internally parallel, but must complete in its entirety before the next phase can begin. A typical example might look something like this:

```
barrier b

in parallel for i ∈ T
    repeat
        // do i's portion of the work of a phase
        b.cycle()
    until terminating condition
```

The cycle method of barrier b (sometimes called wait, next, or even barrier) forces each thread i to wait until *all* threads have reached that same point in their execution. Calling cycle accomplishes two things: it announces to other threads that all work prior to the barrier in the current thread has been completed (this is the *arrival part* of the barrier), and it ensures that all work prior to the barrier in *other* threads has been completed before continuing execution in the current thread (this is the *departure part*). To avoid data races, the arrival part typically includes a release (RW$\parallel$) fence or synchronizing store; the departure part typically ends with an acquire ($\parallel$RW) fence.

The simplest barriers, commonly referred to as *centralized*, employ a small, fixed-size data structure, and consume $\Omega(n)$ time between the arrival of the first thread and the departure of the last. More complex barriers distribute the data structure among the threads, consuming $O(n)$ or $O(n \log n)$ space, but requiring only $\Theta(\log n)$ time.

For any maximum number of threads n, of course, $\log n$ is a constant, and with hardware support it can be a very *small* constant. Some multiprocessors (e.g., the Cray X/XE/Cascade, SGI UV, and IBM Blue Gene series) exploit this observation to provide special constant-time barrier operations (the Blue Gene machines, though, do not have a global address space). With a large number of processors, constant-time hardware barriers can provide a substantial benefit over log-time software barriers.

In effect, barrier hardware performs a global AND operation, setting a flag or asserting a signal once all cores have indicated their arrival. It may also be useful—especially on NRC-NUMA machines, to provide a global OR operation (sometimes known as *Eureka*) that can be used to determine when any *one* of a group of threads has indicated its arrival. Eureka mechanisms are commonly used for parallel search: as soon as one thread has found a desired element (e.g., in its portion of some large data set), the others can stop looking. The principal disadvantage of hardware barriers and eureka mechanisms is that they are difficult to virtualize or share among the dynamically changing processes and threads of a multiprogrammed workload.

The first subsection below presents a particularly elegant formulation of the centralized barrier. The following three subsections present different log-time barriers; a final subsection summarizes their relative advantages.

5.2.1   THE SENSE-REVERSING CENTRALIZED BARRIER

It is tempting to expect a centralized barrier to be easy to write: just initialize a counter to zero, have each thread perform a fetch_and_increment when it arrives, and then spin until the total reaches the number of threads. The tricky part, however, is what to do the second time around. Barriers are meant to be used repeatedly, and without care it is easy to write code in which threads that reach the next barrier "episode" (set of calls to cycle) interfere with threads that have not yet gotten around to leaving the previous episode. Several algorithms that suffer from this bug have actually been published.

Perhaps the cleanest solution is to separate the counter from the spin flag, and to "reverse the sense" of that flag in every barrier episode. Code that embodies this technique appears in Figure 5.1. It is adapted from Hensgen et al. [1988]; Almasi and Gottlieb [1989, p. 445] credit similar code to Isaac Dimitrovsky.

The bottleneck of the centralized barrier is the arrival part: fetch_and_increment operations will serialize, and each can be expected to entail a remote memory access or coherence miss. Departure will also entail $O(n)$ time, but on a globally cache-coherent machine every spinning

```
class barrier
    int count := 0
    const int n := |T|
    bool sense := true
    bool local_sense[T] := { true … }
barrier.cycle():
    bool s := ¬local_sense[self]
    local_sense[self] := s              // each thread toggles its own sense
    if FAI(&count, RW‖) = n−1          // note release ordering
        count.store(0)
        sense.store(s)                  // last thread toggles global sense
    else
        while sense.load() ≠ s;         // spin
    fence(R‖RW)
```

Figure 5.1: The sense-reversing centralized barrier.

thread will have its own cached copy of the sense flag, and post-invalidation refills will generally be able to pipeline or combine, for much lower per-access latency.

5.2.2   SOFTWARE COMBINING

It has long been known that a linear sequence of associative (and, ideally, commutative) operations (a "reduction") can be performed tree-style in logarithmic time [Ladner and Fischer, 1980]. For certain read-modify-write operations (notably fetch_and_add), Kruskal et al. [1988] developed reduction-like hardware support as part of the NYU Ultracomputer project [Gottlieb et al., 1983]. On a machine with a log-depth interconnection network (in which a message from processor i to memory module j goes through a $O(\log p)$ internal switching nodes on a p-processor machine), near-simultaneous requests to the same location *combine* at the switching nodes. For example, if operations FAA(l, a) and FAA(l, b) landed in the same internal queue at about the same point in time, they would be forwarded on as a single FAA(l, a+b) operation. When the result (the original value—call it v) returned (over the same path), it would be split into two responses—v and either $(v+a)$ or $(v+b)$—and returned to the original requesters.

While hardware combining tends not to appear on modern machines, Yew et al. [1987] observed that similar benefits could be achieved with an explicit tree in software. A shared variable that is expected to be the target of multiple concurrent accesses is represented as a tree of variables, with each node in the tree assigned to a different cache line. Threads are divided into groups, with one group assigned to each leaf of the tree. Each thread updates the state in its leaf. If it discovers that it is the last thread in its group to do so, it continues up the tree and updates its parent to reflect the collective updates to the child. Proceeding in this fashion, late-coming threads eventually propagate updates to the root of the tree.

Using a software combining tree, Tang and Yew [1990] showed how to create a log-time barrier. Writes into one tree are used to determine that all threads have reached the barrier; reads

```
type node = record
        const int k                                    // fan-in of this node
        int count := k
        bool sense := false
        node* parent := ...                            // initialized appropriately for tree
class barrier
        bool local_sense[T] := { true ... }
        node* my_leaf[T] := ...                        // pointer to starting node for each thread
        // initialization must create a tree of nodes (each in its own cache line)
        // linked by parent pointers
barrier.cycle():
        fence(RW∥W)
        combining_helper(my_leaf[self], local_sense[self])    // join the barrier
        local_sense[self] := ¬local_sense[self]               // for next barrier
        fence(R∥RW)
combining_helper(node* n, bool my_sense):
        if FAD(&n→count) = 1                           // last thread to reach this node
            if n→parent ≠ null
                combining_helper(n→parent, my_sense)
            n→count.store(n→k)                         // prepare for next barrier episode
            n→sense.store(¬n→sense)                    // release waiting threads
        else
            while n→sense.load() ≠ my_sense;   // spin
```

Figure 5.2: A software combining tree barrier. FAD is fetch_and_decrement.

out of a second are used to allow them to continue. Figure 5.2 shows a variant of this combining tree barrier, as modified by Mellor-Crummey and Scott [1991b] to incorporate sense reversal and to replace the fetch_and_Φ instructions of the second combining tree with simple reads (since no real information is returned).

Simulations by Yew et al. [1987] show that a software combining tree can significantly decrease contention for reduction variables, and Mellor-Crummey and Scott [1991b] confirm this result for barriers. At the same time, the need to perform (typically expensive) fetch_and_Φ operations at each node of the tree induces substantial constant-time overhead. On an NRC-NUMA machine, most of the spins can also be expected to be remote, leading to potentially unacceptable contention. The barriers of the next two subsections tend to work much better in practice, making combining tree barriers mainly a matter of historical interest. This said, the notion of combining—broadly conceived—has proven useful in the construction of a wide range of concurrent data structures. We will return to the concept briefly in Section 5.4.

5.2.3   THE DISSEMINATION BARRIER

Building on earlier work on barriers [Brooks, 1986] and information dissemination [Alon et al., 1987, Han and Finkel, 1988], Hensgen et al. [1988] describe a *dissemination barrier* that re-

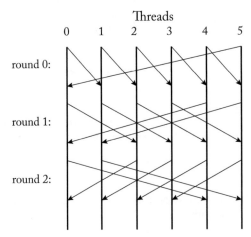

Figure 5.3: Communication pattern for the dissemination barrier (adapted from Hensgen et al. [1988]).

duces barrier latency by eliminating the separation between arrival and departure. The algorithm proceeds through $\lceil \log_2 n \rceil$ (unsynchronized) rounds. In round k, each thread i signals thread $(i + 2^k)$ mod n. The resulting pattern (Figure 5.3), which works for arbitrary n (not just a power of 2), ensures that by the end of the final round every thread has heard—directly or indirectly—from every other thread.

Code for the dissemination barrier appears in Figure 5.4. The algorithm uses alternating sets of variables (chosen via parity) in consecutive barrier episodes, avoiding interference without requiring two separate spins in each round. It also uses sense reversal to avoid resetting variables after every episode. The flags on which each thread spins are statically determined (allowing them to be local even on an NRC-NUMA machine), and no two threads ever spin on the same flag.

Interestingly, while the critical path length of the dissemination barrier is $\lceil \log_2 n \rceil$, the total amount of interconnect traffic (remote writes) is $n \lceil \log_2 n \rceil$. (Space requirements are also $O(n \log n)$.) This is asymptotically larger than the $O(n)$ space and bandwidth of the centralized and combining tree barriers, and may be a problem on machines whose interconnection networks have limited cross-sectional bandwidth.

5.2.4   NON-COMBINING TREE BARRIERS

While the potential cost of fetch_and_Φ operations is an argument against the combining tree barrier, it turns out not to be an argument against tree barriers in general. Hensgen et al. [1988] and Lubachevsky [1989] observe that one can eliminate the need for fetch_and_Φ by choosing the "winner" at each tree node in advance. As in a combining tree, threads in the resulting *tournament barrier* begin at the leaves and move upward, with only one continuing at each level. Instead of

```
const int logN = ⌈log₂ n⌉
type flag_t = record
    bool my_flags[0..1][0..logN−1]
    bool* partner_flags[0..1][0..logN−1]
class barrier
    int parity[T] := { 0 ... }
    bool sense[T] := { true ... }
    flag_t flag_array[T] := ...
        // on an NRC-NUMA machine, flag_array[i] should be local to thread i
        // initially flag_array[i].my_flags[r][k] is false ∀i, r, k
        // if j = (i + 2ᵏ) mod n, then ∀r, k:
        //     flag_array[i].partner_flags[r][k] points to flag_array[j].my_flags[r][k]
barrier.cycle():
    fence(RW‖W)
    flag_t* fp := &flag_array[self]
    int p := parity[self]
    bool s := sense[self]
    for int i in 0..logN−1
        *(fp→partner_flags[p][i]).store(s)
        while fp→my_flags[p][i].load() ≠ s;    // spin
    if p = 1
        sense[self] := ¬s
    parity[self] := 1 − p
    fence(R‖RW)
```

Figure 5.4: The dissemination barrier.

the last arrival, however, it is a *particular* thread (say the one from the left-most child) that always continues upward. Other threads set a flag in the node to let the "winner" know they have arrived. If the winner arrives before its peers, it simply waits. Wakeup can proceed back down the tree, as in the combining tree barrier, or (on a machine with broadcast-based cache coherence) it can use a global flag. With care, the tree can be designed to avoid remote spinning, even on an NRC-NUMA machine, though the obvious way to do so increases space requirements from $O(n)$ to $O(n \log_2 n)$ [Lee, 1990, Mellor-Crummey and Scott, 1991b].

Inspired by experience with tournament barriers, Mellor-Crummey and Scott [1991b] proposed a *static tree barrier* that takes logarithmic time and linear space, spins only on local locations (even on an NRC-NUMA machine), and performs the theoretical minimum number of remote memory accesses ($2n − 2$) on machines that lack broadcast. Unlike a tournament barrier, the static tree barrier associates threads with internal nodes as well as leaves, thereby reducing the overall size of the tree. Each thread signals its parent, which in turn signals *its* parent when it has heard from all of its children.

Code for the static tree barrier appears in Figure 5.5. It incorporates a minor bug fix from Kishore Ramachandran. Each thread is assigned a unique tree node which is linked into an arrival

```
type node = record
     bool parent_sense := false
     bool* parent_ptr
     bool have_child[0..3]                    // for arrival
     bool child_not_ready[0..3]
     bool* child_ptrs[0..1]                   // for departure
     bool dummy                               // pseudodata
```

```
class barrier
     bool sense[T] := true
     node nodes[T]
     // on an NRC-NUMA machine, nodes[i] should be local to thread i
     // in nodes[i]:
     //      have_child[j] = true iff 4i + j + 1 < n
     //      parent_ptr = &nodes[⌊(i − 1)/4⌋].child_not_ready[(i − 1) mod 4],
     //           or &dummy if i = 0
     //      child_ptrs[0] = &nodes[2i + 1].parent_sense, or &dummy if 2i + 1 ≥ n
     //      child_ptrs[1] = &nodes[2i + 2].parent_sense, or &dummy if 2i + 2 ≥ n
     //      initially child_not_ready := have_child
```

```
barrier.cycle():
     fence(RW‖W)
     node* n := &nodes[self]
     bool my_sense := sense[self]
     while n→child_not_ready.load() ≠ { false, false, false, false };      // spin
     n→child_not_ready.store(n→have_child)           // prepare for next episode
     *n→parent_ptr.store(false)                       // let parent know we're ready
     // if not root, wait until parent signals departure:
     if self ≠ 0
          while n→parent_sense.load() ≠ my_sense;          // spin
     // signal children in departure tree:
     *n→child_ptrs[0].store(my_sense)
     *n→child_ptrs[1].store(my_sense)
     sense[self] := ¬my_sense
     fence(R‖RW)
```

Figure 5.5: A static tree barrier with local-spinning tree-based departure.

tree by a parent link and into a wakeup tree by a set of child links. It is useful to think of the trees as separate because their arity may be different. The code shown here uses an arrival fan-in of 4 and a departure fan-out of 2, which worked well in the authors' original (c. 1990) experiments. Assuming that the hardware supports single-byte writes, fan-in of 4 (on a 32-bit machine) or 8 (on a 64-bit machine) allows a thread to use a single-word spin to wait for all of its arrival-tree children simultaneously. Optimal departure fan-out is likely to be machine-dependent. As in the tournament barrier, wakeup on a machine with broadcast-based global cache coherence could profitably be effected with a single global flag.

Table 5.1: Tradeoffs among leading software barriers. Critical path lengths are in remote memory references (assuming broadcast on a CC-NUMA machine); they may not correspond precisely to wall-clock time. Space needs are in words. Constants a and d in the static tree barrier are arrival fan-in and departure fan-out, respectively. Fuzzy barriers are discussed in Section 5.3.1

	central	dissemination	static tree
space needs			
CC-NUMA	n+1	$n + 2n\lceil\log_2 n\rceil$	$4n + 1$
NRC-NUMA			$(5 + d)n$
critical path length			
CC-NUMA	$n + 1$	$\lceil\log_2 n\rceil$	$\lceil\log_a n\rceil + 1$
NRC-NUMA	∞		$\lceil\log_a n\rceil + \lceil\log_d n\rceil$
total remote refs			
CC-NUMA	$n + 1 .. 2n$	$n\lceil\log_2 n\rceil$	n
NRC-NUMA	∞		$2n - 2$
fuzzy barrier suitability	+	−	−
tolerance of changes in n	+	−	−

5.2.5   WHICH BARRIER SHOULD I USE?

Experience suggests that the centralized, dissemination, and static tree barriers are all useful in certain circumstances. Tradeoffs among them are summarized in Table 5.1. Given the cost of remote spinning (and of fetch_and_Φ operations on most machines), the combining tree barrier tends not to be competitive. The tournament barrier (mentioned in Section 5.2.4) likewise has little to recommend it over the static tree barrier.

The centralized barrier has the advantage of simplicity, and tends to outperform all other alternatives when the number of threads is small. It also adapts easily to different numbers of threads. In an application in which the number changes from one barrier episode to another, this advantage may be compelling.

The choice between the dissemination and static tree barriers comes down to a question of architectural features and costs. The dissemination barrier has the shortest critical path, but induces asymptotically more total network traffic. (It is also ill suited to applications that can exploit the "fuzzy" barriers of Section 5.3.1 below.) Given broadcast-based cache coherence, nothing is likely to outperform the static tree barrier, modified to use a global departure flag. In the absence of broadcast, the dissemination barrier will do better on a machine with high cross-sectional bandwidth; otherwise the static tree barrier (with explicit departure tree) is likely to do better. When in doubt, practitioners would be wise to try both and measure their performance.

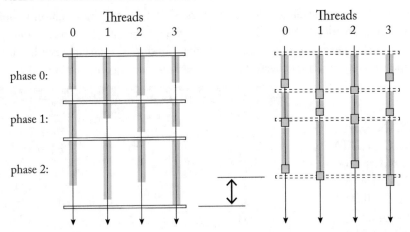

Figure 5.6: Impact of variation across threads in phase execution times, with normal barriers (left) and fuzzy barriers (right). Blue work bars are the same length in each version of the figure. Fuzzy intervals are shown as outlined boxes. With fuzzy barriers, threads can leave the barrier as soon as the last peer has entered its fuzzy interval. Overall performance improvement is shown by the double-headed arrow at center.

5.3   BARRIER EXTENSIONS

5.3.1   FUZZY BARRIERS

One of the principal performance problems associated with barriers is *skew* in thread arrival times, often caused by irregularities in the amount of work performed between barrier episodes. If one thread always does more work than the others, of course, then its arrival will always be delayed, and all of the others will wait. If variations are more normally distributed, we may see the situation illustrated on the left side of Figure 5.6, where the time between barrier episodes is repeatedly determined by a different slowest thread. If $T_{i,r}$ is the time thread i consumes in phase r of the computation, then total execution time is $\sum_r (t_b + \max_{i \in \mathcal{T}} T_{i,r})$, where t_b is the time required by a single barrier episode.

Fortunately, it often turns out that the work performed in one algorithmic phase depends on only *some* of the work performed by peers in previous phases. Imagine, for example, a program that repeatedly updates all the elements of a complex simulated system, collects and logs information about the current state for subsequent (off-line) analysis, and proceeds to the next step. If logs are kept on a per-thread basis, then we can start the next phase of simulation in a fast thread as soon as its peers have finished their local updates: we don't have to wait for them to finish their logging. This observation, due to Gupta [1989], leads to the design of a *fuzzy* barrier, in which arrival and departure are separate operations. The standard idiom

```
class barrier
    int count := 0
    const int n := |T|
    bool sense := true
    bool local_sense[T] := { true ... }

barrier.arrive():
    local_sense[self] := ¬local_sense[self]       // each thread toggles its own sense
    if FAI(&count, RW‖) = n−1                      // note release ordering
        count.store(0)                            // last thread prepares for next episode
        sense.store(local_sense[self])            // and toggles global sense

barrier.depart():
    while sense.load() ≠ local_sense[self];        // spin
    fence(R‖RW)
```

Figure 5.7: Fuzzy variant of the sense-reversing centralized barrier.

```
in parallel for i ∈ T
    repeat
        // do i's portion of the work of a phase
        b.cycle()
    until terminating condition
```

becomes

```
in parallel for i ∈ T
    repeat
        // do i's critical work for this phase
        b.arrive()
        // do i's non-critical work—its fuzzy interval
        b.depart()
    until terminating condition
```

As illustrated on the right side of Figure 5.6, the impact on overall run time can be a dramatic improvement.

A centralized barrier is easily modified to produce a fuzzy variant (Figure 5.7). Unfortunately, none of the logarithmic barriers we have considered has such an obvious fuzzy version. We address this issue in the following subsection.

5.3.2 ADAPTIVE BARRIERS

When all threads arrive at about the same time, tree and dissemination barriers enjoy an asymptotic advantage over the centralized barrier. The latter, however, has an important advantage when thread arrivals are heavily skewed: if all threads but one have already finished their arrival work, the last thread is able recognize this fact in constant time in a centralized barrier. In the other barriers it will almost always require logarithmic time (an exception being the lucky case in which the last-arriving thread just happens to own a node within constant distance of the root of the static tree barrier).

This inability to amortize arrival time is the reason why most log-time barriers do not easily support fuzzy-style separation of their arrival and departure operations. In any fuzzy barrier it is essential that threads wait *only* in the departure operation, and then only for threads that have yet to reach the arrival operation. In the dissemination barrier, no thread knows that all other threads have arrived until the very end of the algorithm. In the tournament and static tree barriers, static synchronization orderings force some threads to wait for their peers before announcing that they have reached the barrier. In all of the algorithms with tree-based departure, threads waiting near the leaves cannot discover that the barrier has been achieved until threads higher in the tree have already noticed this fact.

Among our log-time barriers, only the combining tree appears to offer a way to separate arrival and departure. Each arriving thread works its way up the tree to the top-most unsaturated node (the first at which it was not the last to arrive). If we call this initial traversal the arrive operation, it is easy to see that it involves no spinning. On a machine with broadcast-based global cache coherence, the thread that saturates the root of the tree can flip a sense-reversing flag, on which all threads can spin in their depart operation. On other machines, we can traverse the tree again in depart, but in a slightly different way: this time a thread proceeds upward only if it is the *first* to arrive at a node, rather than the last. Other threads spin in the top-most node in which they were the first to arrive. The thread that reaches the root waits, if necessary, for the last arriving thread, and then flips flags in each of the children of the root. Any thread that finds a flipped flag on its way up the tree knows that it can stop and flip flags in the children of *that* node. Races among threads take a bit of care to get right, but the code can be made to work. Unfortunately, the last thread to call the arrive operation still requires $\Omega(\log n)$ time to realize that the barrier has been achieved.

To address this remaining issue, Gupta and Hill [1989] proposed an *adaptive combining tree*, in which early-arriving threads dynamically modify the structure of the tree so that late-arriving peers are closer to the root. With modest skew in arrival times, the last-arriving thread realizes that the barrier has been achieved in constant time.

The code for this barrier is somewhat complex. The basic idea is illustrated in Figure 5.8. It uses a binary tree. Each thread, in its arrive operation, starts at its (statically assigned) leaf and proceeds upward, stopping at the first node (say, w) that has not yet been visited by any other thread. Having reached w through, say, n, it then modifies the tree so that w's other child, o, is one level closer to the root. Specifically, it changes o's parent to be p (the parent of w) and makes o a child of p. A thread that reaches p through w's sibling, x, will promote o another level, and a later-arriving thread, climbing through o, will traverse fewer levels of the tree than it would have otherwise.

In their paper, Gupta and Hill [1989] present both standard and fuzzy versions of their adaptive combining tree barrier. Unfortunately, both versions retain the remote spins of the

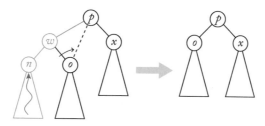

Figure 5.8: Dynamic modification of the arrival tree in an adaptive combining tree barrier.

original (non-adaptive) combining tree. They also employ test_and_set locks to arbitrate access to each tree node. To improve performance—particularly but not exclusively on NRC NUMA machines—Scott and Mellor-Crummey [1994] present versions of the adaptive combining tree barrier (both regular and fuzzy) that spin only on local locations and that adapt the tree in a wait-free fashion, without the need for per-node locks. In the process they also fix several subtle bugs in the earlier algorithms. Shavit and Zemach [2000] generalize the notion of combining to support more general operations than simply "arrive at barrier"; we will return to their work in Section 5.4.

Scott and Mellor-Crummey report mixed performance results for adaptive barriers: if arrival times are skewed across threads, tree adaptation can make a significant difference, both by reducing departure times in the wake of the last arrival and by making fuzzy intervals compatible with a logarithmic critical path. If thread arrival times are very uniform, however, the overhead of adaptation may yield a net loss in performance. As with many other tradeoffs, the break-even point will vary with both the machine and workload.

5.3.3 BARRIER-LIKE CONSTRUCTS

While barriers are the most common form of global (all-thread) synchronization, they are far from the only one. We have already mentioned the "Eureka" operation in Sections 2.3.2 and 5.2. Invoked by a thread that has discovered some desired result (hence the name), it serves to interrupt the thread's peers, allowing them (in the usual case) to stop looking for similar results. Whether supported in hardware or software, the principal challenge for Eureka is to cleanly terminate the peers. The easiest solution is to require each thread to poll for termination periodically, but this can be both awkward and wasteful. More asynchronous solutions require careful integration with the thread library or language run-time system, and are beyond the scope of this lecture.

Many languages (and, more awkwardly, library packages) support *series-parallel* execution, in which an executing thread can launch a collection of children and then wait for their completion. The most common syntax involves a "loop" whose iterations are intended to execute in parallel. While the programmer is typically encouraged to think of such a loop as forking a separate task for each iteration, and joining them at the end, the underlying implementation may just as easily employ a set of preexisting "worker" threads and a barrier at the bottom of the loop. Because the number of workers and the number of tasks cannot in general be assumed to be the same, implementations of series-parallel execution are usually based on scheduler-based synchronization, rather than busy-wait. We will return to this topic in Section 7.4.3.

5.4   COMBINING AS A GENERAL TECHNIQUE

As noted in Section 5.2.2, the software combining tree barrier is a specific instance of a more general combining technique. An important detail, glossed over in that section, is that the first thread to arrive at a node of a combining tree must wait for others to arrive. In a barrier, we know that every thread must participate, exactly once, and so waiting for peers is appropriate.

Suppose, however, that we wish to implement a shared counter object with a fetch_and_add operation. Any thread can invoke fetch_and_add at any time; there is no guarantee that the number of invocations will be balanced across threads, or that any particular peer will invoke an operation within any particular span of time. For objects such as this, we can choose some modest delay interval, pause for that amount of time at each tree node, and continue on up the tree if no peer arrives [Tang and Yew, 1990]. The delays, of course, increase the latency of any individual operation, as does the traversal of a log-depth tree. Under heavy load, this disadvantage is outweighed by the decrease in contention, which leads to higher throughput.

To some extent, one can obtain the best of both worlds: the *combining funnels* of Shavit and Zemach [2000] adapt the width and depth of a combining tree to match the offered load. This adaptation is more general than that of the barrier in Section 5.3.2; combining funnels allow threads to perform their operations at arbitrary times and rates, and to receive individualized return values.

One of the most common applications of combining occurs in *reduction* operations, in which the results of some large number of tasks are "folded" together to produce a single summary value—e.g., the sum, product, maximum, or minimum of the values computed by the tasks. When the combining operation ($+$, $\times$, max, min) is commutative, individual values can be folded into the summary value in any order. When the operation is also associative, values can be folded together in a combining tree. In the simple case, all we care about is the final result, so values flow only from the leaves to the root. In the more general case, in which we want an intermediate result for each individual operation, values also flow back from the root to the leaves; in this latter case, the operation (call it $\oplus$) must be *reversible*: given a, b, and the return value of $v \oplus (a \oplus b)$, where v was the original (usually unknown) value, we must be able to deduce either $v \oplus a$ or $v \oplus b$.

In some cases, combined operations may *eliminate* each other. If we are computing a global sum, for example (without intermediate results), and if "+3" and "−3" operations combine in a tree node, there is really no point in propagating a "+0" message up the tree. A more compelling example occurs in the case of pushes and pops on an abstract stack. Given any stack configuration, a push followed immediately by a pop leaves the same configuration afterward, meaning that as long as the pop returns the value provided by the push, their location (as a pair) in the overall linearization order of the stack is immaterial [Shavit and Touitou, 1997]. While it is tempting to think of a stack as an inherently serial structure, with operations taking turns updating a single top-of-stack "hot spot," elimination makes it possible to build a scalable implementation [Shavit and Zemach, 2000].

In the general case, when threads combine their operations, one proceeds on behalf of both, and the other waits for the first to return. In the more restricted case of elimination, the fact that neither thread must wait raises the possibility of a nonblocking implementation. The *elimination trees* of Shavit and Touitou [1997] can be used to build an "almost stack" that is nonblocking and, while not linearizable, at least *quiescently consistent*: individual pairs of operations may appear to occur out of order, so long as some operation is active, but as soon as the structure is idle, there must be some sequential order that explains the results of all completed operations. More recently, Hendler et al. [2004] have developed an *elimination-backoff stack* that is both nonblocking and linearizable; we will return to it in Section 8.8.

In other recent work, Hendler et al. [2010b] demonstrate that combining can improve performance even in the absence of a log-depth combining tree. Their *flat combining* approach begins with the code of an ordinary sequential data structure, protected by a single lock. To this the authors add a shared nonblocking set of pending requests. When a thread that wants to perform an operation finds that the lock is held, it adds its request to the set and then waits for its completion, rather than waiting for the lock. When the thread that holds the lock completes its operation, it scans the set, combines requests as much as it can, and then applies what remains as a group. Only when the set is empty does it finally release the lock.

During periods of low contention, operations on the data structure occur more or less one at a time, and behavior is equivalent to traditional coarse-grain locking. When contention is high, however, large numbers of requests are completed by a single thread, which then benefits from three important effects. First, when two operations are combined and then applied together, the total amount of work performed is often less than that of performing the operations separately. A push and a pop, for example, when combined, avoid not only any updates to the top-of-stack pointer, but also, in a linked-list implementation, the overhead of allocating and freeing list nodes. In a similar vein, updates to a sorted list can themselves be sorted, and then applied in a single pass. Second, while performing a group of operations, a thread incurs no costly synchronization. Third, because all the operations occur in a single thread, the total number of cache misses is likely to be quite low: parameters may need to be fetched from remote locations, but accesses to the data structure itself are all likely to be local. In follow-on work, Hendler et al. [2010a] demonstrate that additional speedup can be obtained by dividing the work of combining among several parallel worker threads.

CHAPTER 6

Read-mostly Atomicity

In Chapter 4 we considered the topic of busy-wait mutual exclusion, which achieves atomicity by allowing only one thread at a time to execute a critical section. While mutual exclusion is sufficient to ensure atomicity, it is by no means necessary. Any mechanism that satisfies the ordering constraints of Section 3.1.2 will also suffice. In particular, *read-mostly* optimizations exploit the fact that operations can safely execute concurrently, while still maintaining atomicity, if they read shared data without writing it.

Section 6.1 considers the simplest read-mostly optimization: the reader-writer lock, which allows multiple readers to occupy their critical section concurrently, but requires writers (that is, threads that may update shared data, in addition to reading it) to exclude both readers and other writers. To use the "reader path" of a reader-writer lock, a thread must know, at the beginning of the critical section, that it will never attempt to write. Sequence locks, the subject of Section 6.2, relax this restriction by allowing a reader to "upgrade" to writer status if it forces all concurrent readers to back out and retry their critical sections. (Transactional memory, which we will consider in Chapter 9, can be considered a generalization of sequence locks. TM systems typically automate the back-out-and-retry mechanism; sequence locks require the programmer to implement it by hand.) Finally read-copy update (RCU), the subject of Section 6.3, explores an extreme position in which the overhead of synchronization is shifted almost entirely off of readers and onto writers, which are assumed to be quite rare.

6.1 READER-WRITER LOCKS

Reader-writer locks, first suggested by Courtois et al. [1971], relax the constraints of mutual exclusion to permit more than one thread to inspect a shared data structure simultaneously, so long as none of them modifies it. Critical sections are separated into two classes: *writes*, which require exclusive access while modifying protected data, and *reads*, which can be concurrent with one another (though not with writes) because they are known in advance to make no observable changes.

As recognized by Courtois et al., different fairness properties are appropriate for a reader-writer lock, depending on the context in which it is used. A "reader preference" lock minimizes the delay for readers and maximizes total throughput by allowing a newly arriving reader to join a group of current readers even if a writer is already waiting. A "writer preference" lock ensures that updates are seen as soon as possible by requiring readers to wait for any current or waiting writer, even if other threads are currently reading, and even if the writer arrived after some of the readers

```
class rw_lock
    int n := 0
    // low-order bit indicates whether a writer is active;
    // remaining bits are a count of active or waiting readers
    const int WA_flag = 1
    const int RC_inc = 2
    const int base, limit, multiplier = ...        // tuning parameters
```

```
rw_lock.writer_acquire():                    rw_lock.reader_acquire():
    int delay := base                            (void) FAA(&n, RC_inc)
    while ¬CAS(&n, 0, WA_flag);   // spin         while n.load() & WA_flag = 1;   // spin
        pause(delay)                             fence(R‖R)
        delay := min(delay × multiplier, limit)
    fence(R‖RW)
```

```
rw_lock.writer_release():                    rw_lock.reader_release():
    (void) FAA(&n, −WA_flag, RW‖)                 (void) FAA(&n, −RC_inc, R‖)
```

Figure 6.1: A centralized reader-preference reader-writer lock, with exponential backoff for writers.

did. Both of these options permit indefinite postponement and even starvation of non-preferred threads when competition for the lock is high. Though not explicitly recognized by Courtois et al., it is also possible to construct a reader-writer lock (called a "fair" lock below) in which readers wait for any earlier writer and writers wait for any earlier thread of either kind.

The locks of Courtois et al. were based on semaphores, a scheduler-based synchronization mechanism that we will introduce in Section 7.2. In the current chapter we limit ourselves to busy-wait synchronization. Like standard mutual-exclusion locks, reader-writer locks admit a wide range of special-purpose adaptations. Calciu et al. [2013], for example, describe mechanisms to extend the locality-conscious locking of Section 4.5.1 to the reader-writer case.

6.1.1   CENTRALIZED ALGORITHMS

There are many ways to construct a centralized reader-writer lock. We consider three examples here.

Our first example (Figure 6.1) gives preference to readers. It uses an unsigned integer to represent the state of the lock. The lowest bit indicates whether a writer is active; the upper bits contain a count of active or interested readers. When a reader arrives, it increments the reader count (atomically) and waits until there are no active writers. When a writer arrives, it attempts to acquire the lock using CAS. The writer succeeds, and proceeds, only when all bits were clear, indicating that no other writer was active and that no readers were active or interested. Since a reader waits only when a writer is active, and is able to proceed as soon as that one writer finishes, exponential backoff for readers is probably not needed (constant backoff may sometimes be appropriate; we do not consider it here). Since writers may be delayed during the execution of an arbitrary number of critical sections, they use exponential backoff to minimize contention.

The symmetric case—writer preference—appears in Figure 6.2. In this case we must count both active readers (to know when all of them have finished) and interested writers (to know whether a newly arriving reader must wait). We also need to know whether a writer is currently active. Even on a 32-bit machine, a single word still suffices to hold both counts and a Boolean flag. A reader waits until there are no active or waiting writers; a writer waits until there are no active readers. Because writers are unordered (in this particular lock), they use exponential backoff to minimize contention. Readers, on the other hand, can use ticket-style proportional backoff to defer to all waiting writers.

Our third centralized example—a fair reader-writer lock—appears in Figure 6.3. It is patterned after the ticket lock (Section 4.2.2), and is represented by two pairs of counters. Each pair occupies a single word: the upper half of each counts readers; the lower half counts writers. The counters of the request word indicate how many threads have requested the lock. The counters of the completion word indicate how many have already acquired and released it. With arithmetic performed modulo the precision of half-word quantities (and with this number assumed to be significantly larger than the total number of threads), overflow is harmless. Readers spin until all earlier writers have completed. Writers spin until all earlier readers and writers have completed. For both readers and writers, we use the difference between requested and completed numbers of writers to estimate the expected wait time. Depending on how many (multi-)reader episodes are interleaved with these, this estimate may be off by as much as a factor of 2.

In recent work, Brandenburg and Anderson [2010] have introduced a fourth reader-writer lock variant, which they call a *phase-fair* lock. In phase-fair locking, readers and writers alternate, so long as there is a continuing supply of each. A centralized implementation is similar to that of Figure 6.3, but each reader waits for at most one writer, rather than all writers with earlier arrival times. When a writer finishes its critical section, all waiting readers are permitted to proceed. Newly arriving readers are permitted to join the current read session if there are no writers waiting; otherwise they wait until after the first waiting writer. Brandenburg and Anderson demonstrate, both analytically and experimentally, that phase-fair ordering is particularly effective for locks in real-time systems.

6.1.2   QUEUED READER-WRITER LOCKS

Just as centralized mutual exclusion locks—even with backoff—can induce unacceptable contention under heavy load on large machines, so too can centralized reader-writer locks. To reduce the contention problem, Mellor-Crummey and Scott [1991a] showed how to adapt queued spin locks to the reader-writer case. Specifically, they presented reader-preference, writer-preference, and fair reader-writer locks based on the MCS lock (Section 4.3.1). All three variants employed a global counter, which, while never the target of a spin, was nonetheless a source of possible contention.

Krieger et al. [1993] showed how to eliminate the global counter in the fair queued reader-writer lock. A subtle bug in this algorithm was recently documented by Dice et al. [2013], who

```
class rw_lock
    ⟨short, short, bool⟩ n := ⟨0, 0, false⟩
    // high half of word counts active readers; low half counts waiting writers,
    // except for low bit, which indicates whether a writer is active
    const int base, limit, multiplier = ...        // tuning parameters

rw_lock.writer_acquire():
    int delay := base
    loop
        ⟨short ar, short ww, bool aw⟩ := n.load()
        if aw = false and ar = 0                    // no active writer or readers
            if CAS(&n, ⟨ar, ww, false⟩, ⟨ar, ww, true⟩) break
            // else retry
        else if CAS(&n, ⟨ar, ww, aw⟩, ⟨ar, ww+1, aw⟩)
            // I'm registered as waiting
            loop                                    // spin
                ⟨ar, ww, aw⟩ := n.load()
                if aw = false and ar = 0     // no active writer or readers
                    if CAS(&n, ⟨ar, ww, false⟩, ⟨ar, ww−1, true⟩) break outer loop
                pause(delay)                        // exponential backoff
                delay := min(delay × multiplier, limit)
        // else retry
    fence(R‖RW)

rw_lock.writer_release():
    fence(RW‖W)
    short ar, wr; bool aw
    repeat     // fetch-and-phi
        ⟨ar, ww, aw⟩ := n.load()
    until CAS(&n, ⟨ar, ww, aw⟩, ⟨ar, ww, false⟩)

rw_lock.reader_acquire():
    loop
        ⟨short ar, short ww, bool aw⟩ := n.load()
        if ww = 0 and aw = false
            if CAS(&n, ⟨ar, 0, false⟩, ⟨ar+1, 0, false⟩) break
        // else spin
        pause(ww × base)                            // proportional backoff
    fence(R‖R)

rw_lock.reader_release():
    fence(R‖W)
    short ar, ww; bool aw
    repeat     // fetch-and-phi
        ⟨ar, ww, aw⟩ := n.load()
    until CAS(&n, ⟨ar, ww, aw⟩, ⟨ar−1, ww, aw⟩)
```

Figure 6.2: A centralized writer-preference reader-writer lock, with proportional backoff for readers and exponential backoff for writers.

```
class rw_lock
    ⟨short, short⟩ requests := ⟨0, 0⟩
    ⟨short, short⟩ completions := ⟨0, 0⟩
    // top half of each word counts readers; bottom half counts writers
    const int base = ...                    // tuning parameter
rw_lock.writer_acquire():
    short rr, wr, rc, wc
    repeat    // fetch-and-phi increment of writer requests
        ⟨rr, wr⟩ := requests.load()
    until CAS(&requests, ⟨rr, wr⟩, ⟨rr, wr+1⟩)
    loop      // spin
        ⟨rc, wc⟩ := completions.load()
        if rc = rr and wc = wr break      // all previous readers and writers have finished
        pause((wc−wr) × base)
    fence(R‖RW)
rw_lock.writer_release():
    fence(RW‖W)
    short rc, wc
    repeat    // fetch-and-phi increment of writer completions
        ⟨rc, wc⟩ := completions.load()
    until CAS(&completions, ⟨rc, wc⟩, ⟨rc, wc+1⟩)
rw_lock.reader_acquire():
    short rr, wr, rc, wc
    repeat    // fetch-and-phi increment of reader requests
        ⟨rr, wr⟩ := requests.load()
    until CAS(&requests, ⟨rr, wr⟩, ⟨rr+1, wr⟩)
    loop      // spin
        ⟨rc, wc⟩ := completions.load()
        if wc = wr break                  // all previous writers have finished
        pause((wc−wr) × base)
    fence(R‖R)
rw_lock.reader_release():
    fence(R‖W)
    short rc, wc
    repeat    // fetch-and-phi increment of reader completions
        ⟨rc, wc⟩ := completions.load()
    until CAS(&completions, ⟨rc, wc⟩, ⟨rc+1, wc⟩)
```

Figure 6.3: A centralized fair reader-writer lock with (roughly) proportional backoff for both readers and writers. Addition is assumed to be modulo the precision of (unsigned) short integers.

```
type role_t = (reader, active_reader, writer)
type qnode = record
     role_t role
     spin_lock mutex
     bool waiting
     qnode* next
     qnode* prev
class rw_lock
     qnode* tail := null

rw_lock.writer_acquire(qnode* l):
     l→role := writer
     l→waiting := true
     l→next := null
     qnode* pred := swap(&tail, l, W‖)
     if pred ≠ null                              // lock is not free
          pred→next.store(l)
          while l→waiting.load();                // spin
     fence(R‖RW)

rw_lock.writer_release(qnode* l):
     fence(RW‖W)
     qnode* succ := l→next.load()
     if succ = null and CAS(&tail, l, null)
          return                                 // no successor; lock is now free
     repeat succ := l→next.load() until succ ≠ null
     succ→prev.store(null)
     succ→waiting.store(false)
```

Figure 6.4: A fair queued reader-writer lock (declarations and writer routines).

observe that hardware transactional memory (HTM—Chapter 9) can be used both to fix the bug and to significantly simply the code. Dice et al. also provide a pair of software-only fixes; after incorporating one of these, code for the lock of Krieger et al. appears in Figures 6.4 and 6.5.

As in the MCS spin lock, the acquire and release routines expect a qnode argument, which they add to the end of list. Each contiguous group of readers maintains both forward and backward pointers in its segment of the list; segments consisting of writers are singly linked. A reader can begin reading if its predecessor is a reader that is already active, though it must first unblock its successor (if any) if that successor is a waiting reader.

In Mellor-Crummey and Scott's reader-writer locks, as in the MCS spin lock, queue nodes could be allocated in the stack frame of the routine that calls acquire and release. In the lock of Krieger et al., this convention would be unsafe: it is possible for another thread to modify a node an arbitrary amount of time after the node's owner has removed it from the queue. To avoid potential stack corruption, queue nodes must be managed by a dynamic *type-preserving* allocator, as described in the box on page 26.

```
rw_lock.reader_acquire(qnode* l):
    l→role := reader;  l→waiting := true
    l→next := l→prev := null
    qnode* pred := swap(&tail, l, W‖)
    if pred ≠ null                                  // lock is not free
        l→prev.store(pred)
        pred→next.store(l)
        if pred→role.load() ≠ active_reader
            while l→waiting.load();                 // spin
    qnode* succ := l→next.load()
    if succ ≠ null and succ→role.load() = reader
        succ→waiting.store(false)                   // unblock contiguous readers
    l→role.store(active_reader, ‖R)

rw_lock.reader_release(qnode* l):
    fence(R‖W)
    qnode* pred := l→prev.load()
    if pred ≠ null                                  // need to disconnect from predecessor
        pred→mutex.acquire()
        while pred ≠ l→prev.load()
            pred→mutex.release()
            pred := l→prev.load();
            if pred = null break
            pred→mutex.acquire()
        // At this point we hold the mutex of our predecessor, if any.
        if pred ≠ null
            l→mutex.acquire()
            pred→next.store(null)
            qnode* succ := l→next.load()
            if succ = null and ¬CAS(tail, l, pred)
                repeat succ := l→next.load() until succ ≠ null
            if succ ≠ null                          // need to disconnect from successor
                succ→prev.store(pred)
                pred→next.store(succ)
            l→mutex.release()
            pred→mutex.release()
            return
    l→mutex.acquire()
    qnode* succ := l→next.load()
    if succ = null and ¬CAS(tail, l, null)
        repeat succ := l→next.load() until succ ≠ null
    if succ ≠ null                                  // ∃ successor but no predecessor
        bool succ_is_writer := succ→role.load() = writer
        succ→waiting.store(false)
        if ¬succ_is_writer
            succ→prev.store(null)
    l→mutex.release()
```

Figure 6.5: A fair queued reader-writer lock (reader routines).

When a reader finishes its critical section, it removes itself from its doubly-linked group of contiguous readers. To avoid races during the unlink operation, the reader acquires mutex locks on its predecessor's qnode and its own. (These can be very simple, since at most two threads will ever contend for access.) If a reader finds that it is the last member of its reader group, it unblocks its successor, if any. That successor will typically be a writer; the exception to this rule is the subject of the bug repaired by Dice et al..

In their paper on phase-fair locks, Brandenburg and Anderson [2010] also present a queue-based implementation with local-only spinning. As of this writing, their lock and the code of Figures 6.4 and 6.5 appear to be the best all-around performers on medium-sized machines (up to perhaps a few dozen hardware threads). For heavily contended locks on very large machines, Lev et al. [2009b] show how to significantly reduce contention among concurrent readers, at the cost of higher overhead when the thread count is low.

One additional case merits special attention. If reads are much more common than writes, and the total number of threads is not too large, the fastest performance may be achieved with a distributed reader-writer lock consisting of $|\mathcal{T}|$ "reader locks"—one per thread—and one "writer lock" [Hsieh and Weihl, 1992]. The reader_acquire routine simply acquires the reader lock corresponding to the calling thread. The writer_acquire routine acquires first the writer lock and then *all* the reader locks, one at a time. The corresponding release routines release these same component locks (in reverse order in the case of writer_release). Reader locks can be very simple, since they are accessed only by a single reader and the holder of the writer lock. Moreover reader_acquire and reader_release will typically be very fast: assuming reads are more common than writes, the needed reader lock will be unheld and locally cached. The writer operations will be slow of course, and each lock will consume space linear in the number of threads. Linux uses locks of this sort to synchronize some kernel-level operations, with per-core kernel instances playing the role of threads.

6.2   SEQUENCE LOCKS

For read-mostly workloads, reader-writer locks still suffer from two significant limitations. First, a reader must know that it *is* a reader, before it begins its work. A deeply nested conditional that occasionally—but very rarely—needs to modify shared data will force the surrounding critical section to function as a writer every time. Second, a reader must write the metadata of the lock itself to ward off simultaneous writers. Because the write requires exclusive access, it is likely to be a cache miss (in a non-distributed, constant-size lock) when multiple readers are active. Given the cost of a miss, lock overhead can easily dominate the cost of other operations in the critical section.

Sequence locks (*seqlocks*) [Lameter, 2005] address these limitations. A reader is allowed to "change its mind" and become a writer in the middle of a critical section. More significantly, readers only read the lock—they do not update it. In return for these benefits, a reader must be prepared to repeat its critical section if it discovers, at the end, that it has overlapped the execution

```
class seqlock
    int n := 0

int seqlock.reader_start():
    int seq
    repeat    // spin until even
        seq := n.load()
    until seq ≡ 0 mod 2
    fence(R‖R)
    return seq

bool seqlock.reader_validate(int seq):
    return (n.load(R‖) = seq)
```

```
bool seqlock.become_writer(int seq):
    if CAS(&n, seq, seq+1, R‖)
        fence(R‖W)
        return true
    return false

seqlock.writer_acquire():
    int seq
    repeat    // spin
        seq := n.load()
    until seq ≡ 0 mod 2 and CAS(&n, seq, seq+1)
    fence(R‖RW)

seqlock.writer_release():
    int seq := n.load()
    n.store(seq+1, RW‖)
```

Figure 6.6: Centralized implementation of a sequence lock. The CAS instructions in writer_acquire and become_writer need to be write atomic.

of a writer. Moreover, the reader's actions must be simple enough that nothing a writer might do can cause the reader to experience an unrecoverable error—divide by zero, dereference of an invalid pointer, infinite loop, etc. Put another way, seqlocks provide mutual exclusion among writers, but not between readers and writers. Rather, they allow a reader to discover, after the fact, that its execution may not have been valid, and needs to be retried.

A simple, centralized implementation of a sequence lock appears in Figure 6.6. The lock is represented by single integer. An odd value indicates that the lock is held by a writer; an even value indicates that it is not. For writers, the integer behaves like a test-and-test_and_set lock. We assume that writers are rare.

A reader spins until the lock is even, and then proceeds, remembering the value it saw. If it sees the same value in reader_validate, it knows that no writer has been active, and that everything it has read in its critical section is mutually consistent. (We assume that critical sections are short enough—and writers rare enough—that n can never roll over and repeat a value before the reader completes. For real-world integers and critical sections, this is a completely safe assumption.) If a reader sees a different value in validate, however, it knows that it has overlapped a writer and must repeat its critical section.

```
repeat
    int s := SL.reader_start()
    // critical section
until SL.reader_validate(s)
```

It is essential here that the critical section be *idempotent*—harmlessly repeatable, even if a writer has modified data in the middle of the operation, causing the reader to see inconsistent state. In the canonical use case, seqlocks serve in the Linux kernel to protect multi-word time information, which can then be read atomically and consistently. If a reader critical section updates thread-local data (only shared data must be read-only), the idiom shown above can be modified to undo the updates in the case where reader_validate returns false.

If a reader needs to perform a potentially "dangerous" operation (integer divide, pointer dereference, unbounded iteration, memory allocation/deallocation, etc.) within its critical section, the reader_validate method can be called repeatedly (with the same parameter each time). If reader_validate returns true, the upcoming operation is known to be safe (all values read so far are mutually consistent); if it returns false, consistency cannot be guaranteed, and code should branch back to the top of the repeat loop. In the (presumably rare) case where a reader discovers that it really needs to write, it can request a "promotion" with become_writer:

```
loop
    int s := SL.reader_start()
    …
    if unlikely_condition
        if ¬SL.become_writer(s) continue        // return to top of loop
        …
        SL.writer_release()
        break
    else                                        // still reader
        …
        if SL.reader_validate(s) break
```

After becoming a writer, of course, a thread has no further need to validate its reads: it will exit the loop above after calling writer_release.

Unfortunately, because they are inherently speculative, seqlocks induce a host of data races [Boehm, 2012]. Every read of a shared location in a reader critical section will typically race with some write in a writer critical section. These races compound the problem of readers seeing inconsistent state: the absence of synchronization means that updates made by writers may be seen by readers out of order. In a language like C or C++, which forbids data races, a straightforward fix is to label all read locations atomic; this will prevent the compiler from reordering accesses, and cause it to issue special instructions that prevent the hardware from reordering them either. This solution is overly conservative, however: it inhibits reorderings that are clearly acceptable within idempotent read-only critical sections. Boehm [2012] explores the data-race issue in depth, and describes other, less conservative options.

A related ordering issue arises from the fact that readers do not modify the state of a seqlock. Because they only read it, on some machines their accesses will not be globally ordered with respect to writer updates. If threads inspect multiple seqlock-protected data structures, a situation analogous to the IRIW example of Figure 2.4 can occur: threads 2 and 3 see updates to objects X and Y, but thread 2 thinks that the update to X happened first, while thread 3 thinks that the update to Y happened first. To avoid causality loops, writers must update the seqlock using sequentially consistent (write-atomic) synchronizing stores.

Together, the problems of inconsistency and data races are subtle enough that seqlocks are best thought of as a special-purpose technique, to be employed by experts in well constrained circumstances, rather than as a general-purpose form of synchronization. That said, seqlock usage can be safely automated by a compiler that understands the nature of speculation. Dalessandro et al. [2010a] describe a system (in essence, a minimal implementation of transactional memory) in which (1) a global sequence lock serializes all writer transactions, (2) fences and reader_validate calls are inserted automatically where needed, and (3) local state is checkpointed at the beginning of each reader transaction, for restoration on abort. A follow-up paper [Dalessandro et al., 2010c] describes a more concurrent system, in which writer transactions proceed speculatively, and a global sequence lock serializes only the write-back of buffered updates. We will return to the subject of transactional memory in Chapter 9.

6.3   READ-COPY UPDATE

Read-copy update, more commonly known as simply RCU [McKenney, 2004, McKenney et al., 2001], is a synchronization strategy originally developed for use within the operating system kernel, and recently extended to user space as well [Desnoyers et al., 2012]. It attempts to drive the overhead of reader synchronization as close to zero as possible, at the expense of potentially very high overhead for writers. Instances of the strategy typically display the following four main properties:

No shared updates by readers. As in a sequence lock, readers modify no shared metadata before or after performing an operation. While this makes them invisible to writers, it avoids the characteristic cache misses associated with locks. To ensure a consistent view of memory, readers may need to execute R∥R fences on some machines, but these are typically much cheaper than a cache miss.

Single-pointer updates. Writers synchronize with one another explicitly. They make their updates visible to readers by performing a single atomic memory update—typically by "swinging" a pointer (under protection of a lock, or using CAS) to refer to the new version of (some part of) a data structure, rather than to the old version. Readers serialize before or after the writer depending on whether they see this update. (In either case, they see data that was valid at some point after they began their call.)

Unidirectional data traversal. To ensure consistency, readers must never inspect a pointer more than once. To ensure serializability (when it is desired), users must additionally ensure (via program logic) that if writers A and B modify different pointers, and A serializes before B, it is impossible for any reader to see B's update but not A's. The most straightforward way to ensure this is to require all structures to be trees, traversed from the root toward the leaves, and by arranging for writers to replace entire subtrees.

Delayed reclamation of deallocated data. When a writer updates a pointer, readers that have already dereferenced the old version—but have not yet finished their operations—may continue to read old data for some time. Implementations of RCU must therefore provide a (potentially conservative) way for writers to tell that all readers that could still access old data have finished their operations and returned. Only then can the old data's space be reclaimed.

Implementations and applications of RCU vary in many details, and may diverge from the description above if the programmer is able to prove that (application-specific) semantics will not be compromised. We consider relaxations of the single-pointer update and unidirectional traversal properties below. First, though, we consider ways to implement relaxed reclamation and to accommodate, at minimal cost, machines with relaxed memory order.

Grace Periods and Relaxed Reclamation.   In a language and system with automatic garbage collection, the delayed reclamation property is trivial: the normal collector will reclaim old data versions when—and only when—no readers can see them any more. In the more common case of manual memory management, a writer may wait until all readers of old data have completed, and then reclaim space itself. Alternatively, it may append old data to a list for eventual reclamation by some other, bookkeeping thread. The latter option reduces latency for writers, potentially improving performance, but may also increase maximum space usage.

Arguably the biggest differences among RCU implementations concern the "grace period" mechanism used (in the absence of a general-purpose garbage collector) to determine when all old readers have completed. In a nonpreemptive OS kernel (where RCU was first employed), the writer can simply wait until a (voluntary) context switch has occurred in every hardware thread. Perhaps the simplest way to do this is to request migration to each hardware thread in turn: such a request will be honored only after any active reader on the target thread has completed.

More elaborate grace period implementations can be used in more general contexts. Desnoyers et al. [2012, App. D] describe several implementations suitable for user-level applications. Most revolve around a global counter C and a global set S of counters, indexed by thread id. C is monotonically increasing (extensions can accommodate rollover): in the simplest implementation, it is incremented at the end of each write operation. In a partial violation of the no-shared-updates property, S is maintained by readers. Specifically, $S[i]$ will be zero if thread i is not currently executing a reader operation. Otherwise, $S[i]$ will be j if C was j when thread i's current reader operation began. To ensure a grace period has passed (and all old readers have finished), a writer iterates through S, waiting for each element to be either zero or a value greater than or equal to the value just written to C. Assuming that each set element lies in a separate cache line, the updates performed by reader operations will usually be cache hits, with almost no performance impact. Moreover, since each element is updated by only one thread, and the visibility of updates can safely be delayed, no synchronizing instructions are required.

Memory Ordering.   When beginning a read operation with grace periods based on the global counter and set, a thread must update its entry in S using a $\|R$ store, or follow the update with a W∥R fence. At the end of the operation, it must update its entry with a R∥ store, or precede the update with a R∥W fence. When reading a pointer that might have been updated by a writer, a reader must use a R∥ load, or follow the read with a R∥R fence. Among these three forms of ordering, the W∥R case is typically the most expensive (the others will in fact be free on a TSO machine). We can avoid the overhead of W∥R ordering in the common case by requiring the writer to interrupt all potential readers (e.g., with a Posix signal) at the end of a write operation. The signal handler can then "handshake" with the writer, with appropriate memory barriers, thereby ensuring that (a) each reader's update of its element in S is visible to the writer, and (b) the writer's updates to shared data are visible to all readers. Assuming that writers are rare, the cost of the signal handling will be outweighed by the no-longer-required W∥R ordering in the (much more numerous) reader operations.

For writers, the rules are similar to the seqlock case: to avoid causality loops when readers inspect more than one RCU-updatable pointer, writers must use sequentially consistent (write-atomic) synchronizing stores to modify those pointers.

Multi-Write Operations.   The single-pointer update property may become a single-*word* update in data structures that use some other technique for traversal and space management. In many cases, it may also be possible to accommodate multi-word updates, by exploiting application-specific knowledge. In their original paper, for example, McKenney et al. [2001] presented an RCU version of doubly-linked lists, in which a writer must update both forward and backward pointers. The key to their solution is to require that readers search for nodes only in the forward direction. The backward pointers can then be treated as hints that facilitate constant-time deletion. Readers serialize before or after a writer depending on whether they see the change to the forward pointer.

In recent work with a similar flavor, Clements et al. show how to implement an RCU version of balanced binary trees. They begin by noting that rebalancing is a semantically neutral operation, and can be separated from insertion and deletion, allowing the latter to be effected with single-pointer updates to a leaf. Assuming that the tree is linked only from parents to children, they then observe that *rotation*-based rebalancing can be designed to change only a small internal subtree, with a single incoming pointer. We can effect an RCU rotation by creating a new version of the subtree and then swinging the pointer to its root. The portion of the structure above the subtree can safely remain unchanged under the usual single-pointer-update rule. More important, most of the tree fragments *below* the subtree can also safely remain unchanged, and need not be copied.

As an example, suppose we wish to rotate nodes x and z in Figure 6.7. We create new nodes x′ and z′, and initialize their child pointers to refer to trees A, B, and C. Then, with a single store (for atomicity with respect to readers) and under protection of a write lock (for atomicity with respect to other writers), we swing p's left or right child pointer (as appropriate) to refer to z′ instead of x. Trees A, B, and C will remain unchanged. Once a grace period has expired, nodes

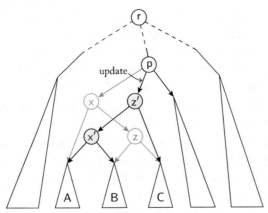

Figure 6.7: Rebalancing of a binary tree via internal subtree replacement (rotation). Adapted from Clements et al. [2012, Fig. 8(b)]. Prior to the replacement, node z is the right child of node x. After the replacement, x′ is the left child of z′.

x and z can be reclaimed. In the meantime, readers that have traveled through x and z will still be able to search correctly down to the fringe of the tree.

In-Place Updates.   As described above, RCU is designed to incur essentially no overhead for readers, at the expense of very high overhead for writers. In some cases, even this property can be relaxed, extending the low-cost case to certain kinds of writers. In the same paper that introduced RCU balanced trees, Clements et al. [2012] observe that trivial updates to page tables—specifically, single-leaf modifications associated with demand page-in—are sufficiently common to be a serious obstacle to scalability on large shared-memory multiprocessors. Their solution is essentially a hybrid of RCU and sequence locks. Major (multi-page) update operations continue to function as RCU writers: they exclude one another in time, install their changes via single-pointer update, and wait for a grace period before reclaiming no-longer-needed space. The page fault interrupt handler, however, functions as an RCU reader. If it needs to modify a page table entry to effect demand page-in, it makes its modifications in place.

This relaxation of the rules introduces a variety of synchronization challenges. For example: a fault handler that overlaps in time with a major update (e.g., an munmap operation that invalidates a broad address range) may end up modifying the about-to-be-reclaimed version of a page table entry, in which case it should not return to the user program as if nothing had gone wrong. If each major update acquires and updates a (per-address-space) sequence lock, however, then the fault handler can check the value of the lock both before and after its operation. If the value has changed, it can retry, using the new version of the data. (Alternatively, if starvation is a concern, it can acquire the lock itself.) Similarly, if fault handlers cannot safely run concurrently with one another (e.g., if they need to modify more than a single word in memory), then they need their own synchronization—perhaps a separate sequence lock in each page table entry. If readers may inspect more than one word that is subject to in-place update, then they, too, may need to inspect such a local sequence lock, and repeat their operation if they see a change. This convention imposes some on the (presumably dominant) read-only code path, but the overhead is still small—in particular, readers still make no updates to shared data.

CHAPTER 7

Synchronization and Scheduling

So far in this lecture, we have emphasized busy-wait synchronization. In the current chapter we turn to mechanisms built on top of a *scheduler*, which multiplexes some collection of cores among a (typically larger) set of threads, switching among them from time to time and—in particular—when the current thread needs to wait for synchronization.

We begin with a brief introduction to scheduling in Section 7.1. We then discuss the oldest (and still most widely used) scheduler-based synchronization mechanism—the semaphore—in Section 7.2. Semaphores have a simple, subroutine-call interface. Many scheduler-based synchronization mechanisms, however, were designed to be embedded in a concurrent programming language, with special, non-procedural syntax. We consider the most important of these—the monitor—in Section 7.3, and others—conditional critical regions, futures, and series-parallel (split-merge) execution—in Section 7.4.

With these mechanisms as background, we return in Section 7.5 to questions surrounding the interaction of user- and kernel-level code: specifically, how to minimize the number of context switches, avoid busy-waiting for threads that are not running, and reduce the demand for kernel resources.

7.1 SCHEDULING

As outlined in Section 1.3, scheduling often occurs at more than one level of a system. The operating system kernel, for example, may multiplex *kernel threads* on top of hardware cores, while a user-level run-time package multiplexes *user threads* on top of the kernel threads. On many machines, the processor itself may schedule multiple *hardware threads* on the pipeline(s) of any given core (in which case the kernel schedules its threads on top of hardware threads, not cores). Library packages (e.g., in Java) may sometimes schedule run-to-completion (unblockable) *tasks* on top of user threads. System-level virtual machine monitors may even multiplex the (virtual) hardware threads seen by guest operating systems on top of some smaller number of physical hardware threads.

Regardless of the level of implementation, we can describe the construction of a scheduler by starting with an overly simple system and progressively adding functionality. The details are somewhat tedious [Scott, 2009, Secs. 8.6, 12.2.4, and 12.3.4]; we outline the basic ideas here in the interest of having "hooks" that we can call in subsequent descriptions of synchronization mechanisms. We begin with *coroutines*—each of which is essentially a stack and a set of registers—and a single core (or kernel thread) that can execute one coroutine at a time. To switch to a different coroutine, the core (or kernel thread) calls an explicit transfer routine, passing as argument a pointer to the *context block* (descriptor) of some other coroutine. The transfer routine (1) pushes all registers other than the stack pointer onto the top of the (current) stack, (2) saves the (updated) stack pointer into the context block of the current coroutine (typically found by examining a global current_thread variable), (3) sets current_thread to the address of the new context block (the argument to transfer), and (4) retrieves a (new) stack pointer from that context block. Because the new coroutine could only have stopped running by calling transfer (and new coroutines are created in such a way that they appear to have just called transfer), the program counter need not change—it will already be at the right instruction. Consequently, the transfer routine simply (5) pops registers from the top of the (new) stack and returns.

On top of coroutines, we implement non-preemptive threads (otherwise known as *run-until-block* or *cooperatively scheduled* threads) by introducing a global *ready list* (often but not always a queue) of runnable-but-not-now-running threads, and a parameterless reschedule routine that pulls a thread off the ready list and transfers to it. To avoid monopolizing resources, a thread should periodically relinquish its core or kernel thread by calling a routine (often named yield) that enqueues it at the tail of the ready list and then calls reschedule. To block for synchronization, the thread can call reschedule after adding itself to some other data structure (other than the ready list), with the expectation that another thread will move it from that structure to the ready list when it is time for it to continue.

The problem with cooperatively scheduled threads, of course, is the need to cooperate—to call yield periodically. At the kernel level, where threads may belong to mutually untrusting applications, this need for cooperation is clearly unacceptable. And even at the user level, it is highly problematic: how do we arrange to yield often enough (and uniformly enough) to ensure fairness and interactivity, but not so often that we spend all of our time in the scheduler? The answer is *preemption*: we arrange for periodic timer interrupts (at the kernel level) or signals (at the user level) and install a handler for the timer that simulates a call to yield in the currently running thread. To avoid races with handlers when accessing the ready list or other scheduler data structures, we temporarily disable interrupts (signals) when executing scheduler operations explicitly.

Given transfer, reschedule/yield, and preemption, we can multiplex concurrent kernel or user threads on a single core or kernel thread. To accommodate true parallelism, we need a separate current_thread variable for each core or kernel thread, and we need one or more spin locks to protect scheduler data structures from simultaneous access by another core or kernel thread. The disabling of interrupts/signals eliminates races between normal execution and timer handlers; spin locks eliminate races among cores or kernel threads. Explicit calls to scheduler routines first disable interrupts (signals) and then acquire the appropriate spin lock(s); handlers simply acquire the lock(s), on the assumption that nested interrupts (signals) are disabled automatically when the first one is delivered.

7.2   SEMAPHORES

Semaphores are the oldest and probably still the most widely used of the scheduler-based synchronization mechanisms. They were introduced by Dijkstra in the mid 1960s [Dijkstra, 1968b]. A semaphore is essentially a non-negative integer with two special operations, P and V.[1] P waits, if necessary, for the semaphore's value to become positive, and then decrements it. V increments the value and, if appropriate, unblocks a thread that is waiting in P. If the initial value of the semaphore is C, is easy to see that $\#P - \#V \leq C$, where $\#P$ is the number of completed P operations and $\#V$ is the number of completed V operations.

---

[1]The names stand for words in Dijkstra's native Dutch: *passeren* (to pass) and *vrijgeven* (to release). English speakers may find it helpful to pretend that P stands for "pause."

Races in the Scheduler

Schedulers are tricky algorithms, with many opportunities for data and (low-level) synchronization races. When implementing (high-level) condition synchronization, for example, the scheduler must generally check a condition and de-schedule the current thread if the condition does not hold. To ensure correctness, we must avoid scenarios in which the corresponding wakeup operation in some other thread falls into the "timing window" between the check of the condition and the operation (typically an enqueue) that makes the waiting thread visible to peers:

```
thread 1:                              thread 2:
    if ¬condition
                                          if ¬Q.empty()
                                              ready_list.enqueue(Q.dequeue())

        Q.enqueue(self)
        reschedule()
```

Here it is important that thread 1 acquire the scheduler spin lock *before* it checks the awaited condition, and hold it through the call to reschedule.

Priority Inversion

The problem addressed by disabling interrupts or signals during scheduler operations is an example of a more general class of problems known as *priority inversion*. Priority inversion occurs when a high priority task (or any sort) preempts a low priority task (or any sort), but is unable to proceed because it needs some resource held by the low priority task. Cast in these terms, a program running above a preemption-based scheduler can be thought of as a low-priority task; an arriving interrupt or signal preempts it, and runs a handler at high priority instead. A spin lock on scheduler data structures ensures atomicity among explicit scheduler operations performed by different cores (or kernel threads), but it cannot provide the same protection between normal execution and interrupt (signal) handlers: a handler that tried to acquire a lock held by the normal code it preempted would end up spinning forever; priority inversion would leave the system deadlocked.

If we let $C = 1$, the semaphore functions as a mutual exclusion lock: P is the acquire operation; V is the release operation. Assuming that the program uses acquire and release correctly (never attempting to release a lock that is not held), the value of the semaphore will always be either 0 (indicating that the lock is held) or 1 (indicating that the lock is free). In this case we say we have a *binary semaphore*. In other cases, a semaphore may represent some general resource of which there are C instances. In this case we say we have a *general* or *counting* semaphore. A thread reserves a resource instance using P; it releases it using V. Within the OS kernel, a semaphore might represent a frame buffer, an optical drive, a physical page of memory, a recurring slot in a time-based communication protocol, or any other resource with a limited, discrete set of instances. Many (though not all) forms of condition synchronization can be captured by the notion of waiting for such a resource.

In Section 1.2 we introduced condition synchronization using the example of a *bounded buffer*, where insert operations would wait, if necessary, for the buffer to become nonfull, and remove operations would wait for it to become nonempty. Code for such a buffer using semaphores appears in Figure 7.1.

Binary and Counting Semaphores

Counting semaphores obviously subsume binary semaphores: given a counting semaphore, we can always choose to initialize it to either 0 or 1, and then perform P and V operations in alternating pairs. As it turns out, we can also build counting semaphores from binary semaphores. Barz [1983] suggests the following implementation:

```
class general_sem
    int count := ...
    binary_sem mutex := 1
    binary_sem gate := min(1, count)

general_sem.P():              general_sem.V():
    gate.P()                      mutex.P()
    mutex.P()                     ++count
    −−count                       if count = 1
    if count > 0                      gate.V()
        gate.V()                  mutex.V()
    mutex.V()
```

The gate binary semaphore serves to serialize P operations. So long as count remains positive, each P "occupies" the gate briefly and then opens it for its successor. A P that drops the count to zero leaves the gate closed; a V that makes it positive again reopens the gate.

The mutual implementability of binary and counting semaphores implies that the two mechanisms are equally powerful. Because their implementations in terms of underlying scheduler primitives are comparable in speed, time, and code size, most systems provide the counting version. A few provide the binary version as well, with extra code to enforce a mutual exclusion-style use pattern: if the program attempts to perform two V operations in a row, a run-time error is announced.

```
class buffer
    const int SIZE = ...
    data buf[SIZE]
    int next_full, next_empty := 0, 0
    semaphore mutex := 1
    semaphore full_slots, empty_slots := 0, SIZE
```

```
buffer.insert(data d):                          buffer.remove():
    empty_slots.P()                                 full_slots.P()
    mutex.P()                                       mutex.P()
    buf[next_empty] := d                            data d := buf[next_full]
    next_empty := (next_empty + 1) mod SIZE         next_full := (next_full + 1) mod SIZE
    mutex.V()                                       mutex.V()
    full_slots.V()                                  empty_slots.V()
                                                    return d
```

Figure 7.1: Implementation of a bounded buffer using semaphores. Semaphore mutex is used to ensure the atomicity of updates to buf, next_full, and next_empty. Semaphores full_slots and empty_slots are used for condition synchronization.

The code for insert and remove is highly symmetric. An initial P operation delays the calling thread until it can claim the desired resource (a full or empty slot). A subsequent brief critical section, protected by the binary mutex semaphore, updates the contents of the buffer and the appropriate index atomically. Finally, a V operation on the complementary condition semaphore indicates the availability of an empty or full slot, and unblocks an appropriate waiting thread, if any.

Given the scheduler infrastructure outlined in Section 7.1, the implementation of semaphores is straightforward. Each semaphore is represented internally by an integer counter and a queue for waiting threads. The P operation disables signals and acquires the scheduler spin lock. It then checks to see whether the counter is positive. If so, it decrements it; if not, it adds itself to the queue of waiting threads and calls reschedule. Either way (immediately or after subsequent wakeup), it releases the scheduler lock, reenables signals, and returns. (Note that the reschedule operation, if called, will release the scheduler lock and reenable signals after pulling a new thread off the ready list. That thread will, in turn, reacquire the lock and disable signals before calling back into the scheduler.) The V operation disables signals, acquires the scheduler lock, and checks to see whether the queue of waiting threads is empty. If so, it moves a thread from that queue to the ready list; if not, it increments the counter. Finally, it releases the scheduler lock, reenables signals, and returns.

With very similar implementation techniques, we can implement native support for scheduler-based reader-writer locks (we could also build them on top of semaphores). Modest changes to the internal representation (protected, of course, by the disabling of signals and the scheduler lock), would lead to fair, reader-preference, or writer-preference versions. In a similar vein, while we have described the behavior of counting semaphores in terms of a "queue" of waiting threads (suggesting FIFO ordering), the choice of thread to resume in V could just as easily be arbitrary, randomized, or based on some notion of priority.

7.3 MONITORS

To a large extent, the enduring popularity of semaphores can be attributed to their simple subroutine-call interface: implemented by a run-time library or operating system, they can be used in almost any language. At the same time, the subroutine-call interface can be seen as a liability. To start with, while mutual exclusion constitutes the most common use case for semaphores, the syntactic independence of P and V operations makes it easy to omit one or the other accidentally—especially in the presence of deeply nested conditions, break and return statements, or exceptions. This problem can be addressed fairly easily by adding syntax in which lock acquisition introduces a nested scope, e.g.:

```
with_lock(L) {
    // ...
}
```

The compiler can ensure that the lock is released on any exit from the critical section, including those that occur via break, return, or exception. In languages like C++, which provide a *destructor* mechanism for objects, a similar effect can be achieved without extending language syntax:

```
{std::lock_guard<std::mutex> _(L);
    // ...
}
```

This construct declares a dummy object (here named simply with an underscore) of class lock_guard. The constructor for this object takes a parameter L of class mutex, and calls its acquire method. The destructor for the unnamed object, which will be called automatically on any exit from the scope, calls L's release method. Both mutex and lock_guard are defined in the C++ standard library.

While scope-based critical sections help to solve the problem of missing acquire and release calls, the association between a lock and the data it protects is still entirely a matter of convention. Critical sections on a given lock may be widely scattered through the text of a program, and condition synchronization remains entirely ad hoc.

To address these limitations, Dijkstra [1972], Brinch Hansen [1973], and Hoare [1974] developed a language-level synchronization mechanism known as the *monitor*. In essence, a monitor is a data abstraction (a module or class) whose methods (often called *entries*) are automatically translated into critical sections on an implicit per-monitor lock. Since fields of the monitor are visible only within its methods, language semantics ensure that the state of the abstraction will be read or written only when holding the monitor lock. To accommodate condition synchronization, monitors also provide *condition variables*. A thread that needs to wait for a condition within the monitor executes a wait operation on a condition variable; a thread that has made a condition true performs a signal operation to awaken a waiting thread. Unlike semaphores, which count the difference in the number of P and V operations over time, condition variables contain only a queue of waiting threads: if a signal operation occurs when no threads are waiting, the operation has no effect.

Over the pasts 40 years, monitors have been incorporated into dozens of programming languages. Historically, Concurrent Pascal [Brinch Hansen, 1975], Modula [Wirth, 1977], and Mesa [Lampson and Redell, 1980] were probably the most influential. Today, Java [Goetz et al., 2006] is probably the most widely used. There have also been occasional attempts to devise a library interface for monitors, but these have tended to be less successful: the idea depends quite heavily on integration into a language's syntax and type system.

Details of monitor semantics vary from one language to another. In the first subsection below we consider the classic definition by Hoare. Though it is not followed precisely (to the best of the author's knowledge) by any particular language, it is the standard against which all other variants are compared. The following two subsections consider the two most significant areas of disagreement among extant monitor variants. The final subsection describes the variant found in Java.

7.3.1   HOARE MONITORS

As originally defined by Hoare [1974], a monitor can be represented pictorially as shown in Figure 7.2. Threads queue for entry at the left and exit at the right. Only one is allowed "inside the box" at any given time. When a thread performs a wait operation, it steps out of the box and into a condition queue. When a thread performs a signal operation, it checks to see whether any thread is waiting on the associated condition queue. If so, it steps out of the box and into the *urgent* queue, and the thread at the head of the condition queue steps in. When a thread exits the monitor, the implementation allows a thread to enter from the urgent queue or, if that queue is empty, from the entry queue. If both threads are empty, the monitor is unlocked.

Some implementations of monitors dispense with the urgent queue, and move signalers back to the entry queue instead. Some implementations also relax the ordering constraints on the various queues, and unblock, when appropriate, an arbitrary, random, or high-priority thread instead of the first in line.

We can use Hoare monitors to implement a bounded buffer as shown in Figure 7.3. It can be instructive to compare it to the semaphore version in Figure 7.1. Where Figure 7.1 uses a binary semaphore for mutual exclusion, Figure 7.3 relies on the implicit mutual exclusion of the monitor lock. More significantly, where Figure 7.1 uses a pair of general semaphores for condition synchronization, and performs a P operation at the beginning of every call to insert or remove, Figure 7.3 inspects integer variable full_slots, and waits *only* when its value indicates that the desired condition does not hold. This difference reflects the fact that semaphores "remember" an excess of V operations, but monitor condition variables do not remember an excess of signals.

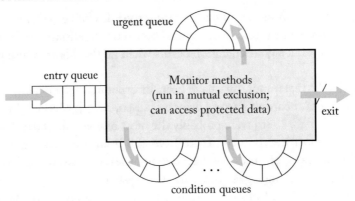

Figure 7.2: A Hoare monitor. Only one thread is permitted "inside the box" at any given time.

```
monitor buffer
      const int SIZE = ...
      data buf[SIZE]
      int next_full, next_empty := 0, 0
      int full_slots := 0
      condition full_slot, empty_slot
```

```
entry insert(data d):                          entry remove():
      if full_slots = SIZE                            if full_slots = 0
            empty_slot.wait()                               full_slot.wait()
      buf[next_empty] := d                          data d := buf[next_full]
      next_empty := (next_empty + 1) mod SIZE       next_full := (next_full + 1) mod SIZE
      ++full_slots                                 −−full_slots
      full_slot.signal()                           empty_slot.signal()
                                                   return d
```

Figure 7.3: Implementation of a bounded buffer as a Hoare monitor. Threads wait on condition variables full_slot and empty_slot only when the associated condition does not currently hold.

Hoare's original paper contains a concise definition of monitors in terms of semaphores. It also notes that a monitor can be used, trivially, to implement a general semaphore, thus proving that the two notations are equally powerful. The principal advantage of the monitor is data abstraction: protected variables (fields) cannot be accessed outside the monitor that contains them, and mutual exclusion on calls to monitor methods is guaranteed implicitly, without the need for explicit acquire (P) and release (V) operations.

To argue that a concurrent data abstraction has been implemented correctly with a monitor, one typically specifies a *monitor invariant*—a logical property of the protected data that must be true whenever the monitor is unoccupied, and whenever the thread inside is changed. Clearly a monitor invariant must always be true initially. It must similarly be true on monitor exit, at every call to wait, and (in a Hoare monitor) on every call to signal. In our bounded buffer example, an appropriate invariant might be that the full_slots variable correctly indicates the number of items in the buffer, and that these items reside in slots next_full through next_empty−1 (mod SIZE) of buf. (We also know that threads will be waiting on the full_slot condition variable only if full_slots is 0, and on the empty_slot condition variable only if full_slots is SIZE, but since the thread queues are generally hidden, this information may not be included in the invariant.)

7.3.2   SIGNAL SEMANTICS

In a Hoare monitor, the thread performing a signal operation (the "signaler") steps out of the monitor and into the urgent queue, so the thread performing a wait operation (the "signalee") can execute immediately. The argument for this behavior is that it allows the signalee to assume that the condition discovered by the signaler still holds. If the signaler were permitted to continue, it might (intentionally or inadvertently) change the state of the monitor before leaving in such a way that the condition no longer held. If any other thread could enter the monitor *between* the signaler and the signalee, there would be no way to ensure the condition.

Unfortunately, immediate resumption of the signalee implies that the signaler must be blocked and then unblocked before continuing execution, entailing two arguably unnecessary context switches of the core on which it is executing. An immediate switch also requires that the signaler reestablish the monitor invariant before the call to signal; it cannot defer this action to code between the signal and the end of the entry. In an attempt to reduce the overhead of context switches, and also to relax reestablishment of the monitor invariant, the designers of Mesa [Lampson and Redell, 1980] chose to define signals as "hints" instead of "absolutes." Mesa dispenses with the urgent queue, and allows a signaling thread to continue execution, switching only when it reaches the end of its entry. Code that would be written

```
if ¬condition
    cvar.wait()
```

in a Hoare monitor would be

```
while ¬condition
    cvar.wait()
```

in a Mesa monitor. This change is certainly not onerous. It is also consistent with the notion of *covering* conditions, discussed in the box on page 112. Most modern implementations of monitors adopt Mesa semantics for signals.

As it turns out, many algorithms (including our bounded buffer) naturally place signal operations only at the ends of entries. A few languages—notably Concurrent Pascal—have *required* this positioning of signals, thereby maintaining the semantics of signals as absolutes while avoiding any extra context switches for immediate transfer to the signalee.

7.3.3   NESTED MONITOR CALLS

A second major difference among monitor implementations concerns behavior in the event of nested calls. Suppose a thread calls entry E of monitor M1, which in turn calls entry F of monitor M2, and the code in F then waits on a condition variable. Clearly M2's monitor lock will be released. But what about M1? If we leave it locked, the program will deadlock if the only way for another thread to reach the necessary signal in M2 is through M1. If we unlock it, however, then the waiting thread in M2 will need to reacquire it when it wakes up, and we may deadlock if some other thread is holding M1's lock at that time—especially if that thread can't release M1's lock without making a nested call to M2.

A possible solution, suggested by Wettstein [1978], is to release the outer monitor when waiting, dictate that signals are only hints, and arrange for a re-awakened lock to re-acquire locks *from the outside in*—i.e., first on M1 and then on M2. This strategy is deadlock free so long as the programmer takes care to ensure that nested monitor calls always occur in the same order (i.e., always from M1 to M2, and never from M2 to M1).

Unfortunately, any scheme in which a nested wait releases the locks on outer monitors will require the programmer to restore the monitor invariant not only on monitor exit, wait, and possibly signal, but also whenever calling an entry of another monitor that may wait—or a subroutine that may call such an entry indirectly. The designers of most languages—Java among them—have concluded that this requirement constitutes an unacceptable burden, and have opted to leave the outer monitor locked.

7.3.4   JAVA MONITORS

The original design of Java—still present in the core of the language—combines scope-based critical sections with a simplified form of monitors. Each object is associated, implicitly, with a mutex

---

Covering Conditions and Cascading Signals

Several languages, including Mesa and Java, provide a broadcast or signalAll operation that awakens *all* threads waiting on a condition. Such an operation finds obvious use in cases where all threads should continue: it makes it trivial, for example, to implement a monitor-based barrier. Broadcast can also be used in programs where the conditions on which threads may wait cannot easily be statically enumerated. Consider, for example, a concurrent set that provides a remove(v) method that waits until v is a member of the set. Absent a separate condition variable for every possible value of v, waiting threads must share a *covering condition*. When such threads may be present, a thread performing an insert operation must broadcast the covering condition, awakening all threads. Since at most one thread will continue in this case, while the rest discover that they must wait again, covering conditions can lead to very high overhead, and must be used with care.

In the absence of broadcast operations, one can employ a *cascading signal* idiom, in which only one thread is initially awoken. If unable to proceed, it explicitly re-signals the condition variable before waiting on it again. Unfortunately, this idiom requires both FIFO ordering of waiting threads (which some systems may not provide) and some mechanism to avoid an infinite loop of signals in the case where no waiting thread is able to proceed.

lock and a single condition variable. A critical section on the lock is specified by a `synchronized` block that names the object:

```
synchronized (my_object) {
    // critical section
}
```

As a form of syntactic sugar, the declaration of a class method can be prefaced with `synchronized`, in which case its body behaves as if surrounded by `synchronized (this) { ... }`. A class whose methods are all `synchronized` functions as a monitor.

Within a `synchronized` method or block, a thread can block for condition synchronization by executing the `wait` method, which all objects inherit from the root class `Object`; it can unblock a waiting peer by executing `notify`. If threads need to wait for more than one condition associated with some abstraction (as they do in our bounded buffer), one must either restructure the code in such a way that each condition is awaited in a different object, or else use some single object's one condition variable to cover all the alternatives. To unblock *all* threads waiting in a given object, one can execute `notifyAll`.

C# provides mechanisms similar to those of core Java. Its `lock` statement is analogous to `synchronized`, and conditions are awaited and signaled with `Wait`, `Pulse`, and `PulseAll`.

The Java 5 revision of the language, released in 2004, introduced a new library-based interface to monitors. Its `Lock` class (with a capital 'L') has explicit `lock` (acquire) and `unlock` (release) methods. These can be used for hand-over-hand locking (Section 3.1.2) and other techniques that cannot easily be captured with scope-based critical sections. Locks can also have an arbitrary number of associated condition variables, eliminating many unnecessary uses of covering conditions. Unfortunately, the library-based interface makes programs somewhat awkward. There is no equivalent of the `synchronized` label on methods, and the Lock-based equivalent of a `synchronized` block looks like this:

```
Lock l = ...;
...
l.lock();
try {
    // critical section
} finally {
    // l.unlock();
}
```

The C# standard library also provides more general synchronization mechanisms, via its `WaitHandle` objects, but these are operating-system specific, and may behave differently on different systems.

```
class buffer
    const int SIZE = ...
    data buf[SIZE]
    int next_full, next_empty := 0, 0
```

```
buffer.insert(data d):                    buffer.remove():
    region when full_slots < SIZE             data d
        buf[next_empty] := d                  region when full_slots > 0
        next_empty := (next_empty + 1)            d := buf[next_full]
                        mod SIZE                  next_full := (next_full + 1) mod SIZE
                                              return d
```

Figure 7.4: Implementation of a bounded buffer using conditional critical regions. Here we have assumed that regions are with respect to the current object (this) unless otherwise specified.

7.4   OTHER LANGUAGE MECHANISMS

7.4.1   CONDITIONAL CRITICAL REGIONS

In Section 1.2 we introduced condition synchronization in the form of a hypothetical await *condition* statement. One potential complaint with both semaphores and monitors is that they require explicit variables as "stand-ins" for Boolean conditions. This observation led Brinch Hansen [1973, Sec. 3.4.3] to propose a notation—the *conditional critical region*—in which awaited conditions could be specified directly. Critical conditional regions appear in several languages, including Edison [Brinch Hansen, 1981] and Ada 95 [Intermetrics, 1995, Secs. II.9 and 9.1]. Syntax is generally some variant on

```
region protected_variable when condition
    // ...
```

As in a Java synchronized block, the protected_variable specifies an object whose implicit lock is to be acquired. Some languages allow the programmer to specify a *list* of objects, in which case their locks are acquired in some canonical order (to avoid deadlock). Significantly, the when clause (also known as a *guard*) can appear only at the *beginning* of the critical section. The intent is that the enclosed code execute atomically at some point in time where the specified condition is true. This convention avoids the issue of monitor signal semantics, but leaves the issue of nested calls.

Figure 7.4 uses conditional critical sections to implement a bounded buffer. The code is arguably more natural than the semaphore (Figure 7.1) or monitor (Figure 7.3) versions, but raises a crucial implementation question: when and how are the guards evaluated?

With no restrictions on the conditions tested by guards, we are faced with the prospect, when one thread leaves a region, of context switching into every other thread that is waiting to enter a region on the same object, so that each can evaluate its own condition in its own referencing environment. With a bit more sophistication, we may be able to determine—statically or at run time—the set of variables on which a condition depends, and only switch into a thread when one of these has changed value (raising the possibility that the condition may now be true). This optimization turns out to be natural in the context of transactional memory; we will return to it in Section 9.3.2. Depending on the cost of tracking writes, it may be cheaper in practice than resuming every thread on every region exit, but worst-case overhead remains significant.

Another cost-reduction strategy, originally proposed by Kessels [1977] and adopted (in essence) by Ada, is to require conditions to depend *only* on the state of the lockable object (never on the parameters passed to methods), and to list these conditions explicitly in the object's declaration. These rules allow the implementation to associate each condition with an implicit queue of waiting threads, and to evaluate it in a generic context, without restoring the referencing environment of any particular thread. When one thread leaves a region, each condition can be evaluated exactly once, and a corresponding thread resumed if the condition is true.

As noted in Section 5.1, it is important that tests of a condition not race with updates to the variables on which the condition depends. This property, too, can be ensured by allowing conditions to depend only on the state of the lockable object—and perhaps also on parameters passed by value, which are inaccessible to other threads.

7.4.2 FUTURES

Futures, first proposed by Halstead [1985] for the Multilisp dialect of Scheme, exploit the observation that function arguments, in most languages, are evaluated before they are passed, but may not actually be *used* by the caller for some time. In Multilisp, any expression—but most commonly a function argument—can be enclosed in a future construct:

```
(future expression)
```

Evaluation of the expression may then proceed in parallel with continued execution in the caller, up until the point (if any) at which the caller actually needs the value of the expression.

Futures embody synchronization in the sense that evaluation of the enclosed expression will not begin until execution in the parent thread reaches the point at which the future appears, and execution in the parent thread will not proceed beyond the point where the value is needed until evaluation has completed. Using futures, the key recursive step in quicksort might be written as follows:

```
(append (future (sort elements less than or equal to pivot))
        (list pivot)
        (future (sort elements greater than pivot)))
```

In general, a future and the continuation of its caller need to be independent, up to the point where the value of the future is needed. If the threads executing the future and the continuation share a data or synchronization race, behavior of the program may be nondeterministic or even undefined. As recognized by Halstead, futures are thus particularly appealing in the purely functional subset of Scheme, where the lack of side effects means that an expression will always evaluate to the same value in a given context.

Some thread libraries provide futures outside the language core—typically as a generic (polymorphic) object whose constructor accepts a closure (a subroutine and its parameters) and whose get method can be used to retrieve the computed value (waiting for it if necessary). In Java, given a Callable<T> object o, the code

```
T val = o.call();
```

can be replaced by

```
Future<T> f = new Future<T>(c);
f.run();
...
T val = f.get();
```

Because Java is not a functional language, the programmer must exercise special care to ensure that a future will execute safely. Welc et al. [2005] have proposed that futures be made safe in *all* cases, using an implementation reminiscent of transactional memory. Specifically, they use *multiversioning* to ensure that a future does not observe changes made by the continuation of its caller, and speculation in the caller to force it to start over if it fails to observe a change made by the future.

7.4.3 SERIES-PARALLEL EXECUTION

At the beginning of Chapter 5, we noted that many spin-based mechanisms for condition synchronization can be rewritten, straightforwardly, to use scheduler-based synchronization instead. Examples include flags, barriers, and reader-writer locks. Any place a thread might spin for a (statically known) condition, we can lock out signals or interrupts, grab the scheduler lock, move the current thread to a condition queue, and call reschedule instead.

Scheduler-based implementations are also commonly used for *series-parallel* execution, mentioned briefly in Section 5.3.3. In Cilk [Blumofe et al., 1995, Frigo et al., 1998], for example, a multi-phase application might look something like this:

```
do {
    for (i = 0; i < n; i++) {
        spawn work(i);
    }
    sync;   // wait for all children to complete
} while ¬terminating condition
```

Conceptually, this code suggests that the main thread create ("fork") *n* threads at the top of each do loop iteration, and "join" them at the bottom. The Cilk runtime system, however, is designed to make spawn and sync as inexpensive as possible. Concise descriptions of the tasks are placed into a "work-stealing queue" [Blumofe et al., 1995] from which they are farmed out to a collection of preexisting worker threads. Similar schedulers are used in a variety of other languages as well. Source code syntax may vary, of course. X10 [Charles et al., 2005], for example, replaces spawn and sync with async and finish.

Many languages (including the more recent Cilk++) include a "parallel for" loop whose iterations proceed logically in parallel. An implicit sync causes execution of the main program to wait for all iterations to complete before proceeding with whatever comes after the loop. Similar functionality can be added to existing languages in the form of *annotations* on sequential loops. OpenMP [Chandra et al., 2001], in particular, defines a set of compiler- or preprocessor-based pragmas that can be used to parallelize loops in C and Fortran. Like threads executing the same phase of a barrier-based application, iterations of a parallel loop must generally be free of data races. If occasional conflicts are allowed, they must be resolved using other synchronization.

In a very different vein, Fortran 95 and its descendants provide a forall loop whose iterations are heavily synchronized. Code like the following

```
forall (i=1:n)
    A[i] = expr1
    B[i] = expr2
    C[i] = expr3
end forall
```

contains (from a semantic perspective) a host of implicit barriers: All instances of *expr1* are evaluated first, then all writes are performed to A, then all instances of *expr2* are evaluated, followed by all writes to B, and so forth. A good compiler will elide any barriers it can prove to be unneeded.

In contrast to unstructured fork-join parallelism, in which a thread may be created—or its completion awaited—at any time, series-parallel programs always generate properly nested groups of tasks. The difference is illustrated in Figure 7.5. With fork and join (a), tasks may join their parent out of order, join with a task other than the parent, or terminate without joining at all. With spawn and sync (b), the parent launches tasks one at a time, but rejoins them as a group. In *split-merge* parallelism (c), we think of the parent as dividing into a collection of children, all at once, and then merging together again later. While less flexible, series-parallel execution leads to clearer source code structure. Assuming that tasks do not conflict with each other, there is also an obvious equivalence to serial execution. For debugging purposes, series-parallel semantics may even facilitate the construction of efficient *race detection* tools, which serve to identify unintended conflicts [Raman et al., 2012].

Recognizing the host of different patterns in which parallel threads may synchronize, Shirako et al. [2008] have developed a barrier generalization known as *phasers*. Threads can join (*register with*) or leave a phaser dynamically, and can participate as *signalers*, *waiters*, or both. Their signal and wait operations can be separated by other code to effect a fuzzy barrier (Section 5.3.1). Threads can also, as a group, specify a statement to be executed, atomically, as part of a phaser episode. Finally, and perhaps most importantly, a thread that is registered with multiple phasers can signal or wait at all of them together when it performs a signal or wait operation. This capability facilitates the management of *stencil* applications, in which a thread synchronizes with its neighbors at the end of each phase, but not with other threads. Neighbor-only synchronization is also supported, in a more limited fashion, by the *topological barriers* of Scott and Michael [1996].

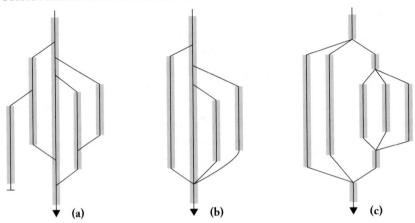

Figure 7.5: Parallel task graphs for programs based on (a) fork and join, (b) spawn and sync, and (c) parallel enumeration (split-merge).

In the message-passing word, barrier-like operations are supported by the *collective communication* primitives of systems like MPI [Bruck et al., 1995], but these are beyond the scope of this lecture.

7.5   KERNEL/USER INTERACTIONS

Throughout this chapter, we have focused on scheduling as a means of implementing synchronization. It is also, however, a means of sharing some limited number of cores or kernel threads among a (usually larger) number of kernel or user threads. Fair sharing requires that user-level applications cross into the kernel when switching between kernel threads, introducing nontrivial overheads. Inopportune preemption—e.g., during a critical section—may introduce greater overheads. Resources required for synchronization (e.g., headers for queues of waiting threads) may constitute a significant burden for systems with large numbers of synchronization objects. The subsections below address these issues in turn.

7.5.1   CONTEXT SWITCHING OVERHEAD

As we noted in Section 1.3, spinning is generally preferable to blocking when the expected wait time is less than twice the context-switching overhead. When switching between kernel threads in user space, this overhead can easily be hundreds of cycles, due to the cost of crossing into kernel mode and back. Some 30 years ago, Ousterhout [1982] suggested that a user-level thread that is unable to acquire a lock (or that is waiting for a currently-false condition) spin for some modest amount of time before invoking the kernel-level scheduler, in the hope that it might avoid the context switch. Many widely used locks today, including most implementations of pthread locks and Java synchronized blocks, use *spin-then-block* techniques. Karlin et al. [1991] provide a detailed analysis of how long to spin before blocking; they demonstrate the value of dynamic adaptation, and prove bounds on worst-case performance.

Of course, spinning in user space requires that the state of the lock be visible in user-level memory. This is natural for locking among threads of the same application, but not for threads in separate address spaces. Traditional interprocess locks, such as Unix System V semaphores, are implemented entirely within the kernel. The semget system call returns an opaque handle, which can be passed to P and V (semop). Spinning before blocking is simply not an option. Sadly, this implies an overhead of hundreds of cycles to acquire even an uncontended lock.

To accommodate fast user-level synchronization *across* address spaces, Franke and Russell [2002] introduced the notion of a *futex*—a fast userspace mutex. Now widely used in Linux and related operating systems, futexes require at least one page of memory to be mapped into the address spaces of all processes sharing a lock. The futex syscalls are intended for use only by user-space thread libraries, and are available only in assembly language. They manipulate data structures in shared memory, and trap into the kernel only when required. In any thread that *does* need to block, the kernel-level context block is modified to indicate the location on which the thread is waiting; this allows the OS to recover cleanly if a process misuses the interface, or terminates unexpectedly. In a similar vein, Solaris provides lwp_park and lwp_unpark syscalls that allow a thread to be descheduled and rescheduled explicitly. Oracle's adaptive mutex library uses these calls to build an enhanced analog of futex. Johnson et al. [2010] discuss the behavior of these mechanisms under a variety of workloads, and present adaptive techniques to control not only the duration of spinning, but also the number of threads that are permitted to spin at any given time.

7.5.2   PREEMPTION AND CONVOYS

Authors of parallel programs have known for decades that performance can suffer badly if a thread is preempted while holding a lock; this is sometimes referred to as *inopportune preemption*. If other threads attempt to acquire the lock, they will need to wait until the lock holder is rescheduled and can complete its critical section. If threads block while waiting, there is reason to hope that the preempted thread will acquire one of the freed-up kernel threads or cores, but if threads spin while waiting instead, an entire scheduling quantum (or more) may expire before the lock holder gets to run again.

Inopportune preemption can generate contention even in otherwise well-balanced programs. Blasgen et al. [1979] describe what they call the *convoy phenomenon*. Suppose that every thread in an application attempts to execute a brief critical section on lock L on a regular but infrequent basis—say once per millisecond. Normally we would not expect the lock to be a bottleneck. But if lock holder T is preempted for more than a millisecond, every other thread may be waiting for L by the time T runs again, and the lock may *become* a bottleneck. Worse, if threads tend to follow similar code paths, their actions coming out of back-to-back critical sections may generate a storm of cache coherence traffic, and they may end up contending for whatever lock is accessed next in program order. Over time the contention may abate, as execution histories drift apart, but as soon as a lock holder is preempted again, the pattern may repeat.

To address the convoy phenomenon, Edler et al. [1988] introduced the notion of *temporary non-preemption*. In a region of memory shared between them, the user thread and the kernel maintain a pair of flags. The first flag is written by the user thread and read by the kernel: it indicates that the user wishes not to be preempted. The second flag is written by the kernel and read by the user thread: it indicates that the kernel wanted to preempt the thread, but refrained because the first flag was set. Ignoring memory ordering constraints, a test_and_set spin lock might use these flags as follows:

```
class lock
    bool f := false

lock.acquire():                              lock.release():
    do_not_preempt_me := true                    f := false
    while ¬TAS(&f)                               do_not_preept_me := false
        do_not_preempt_me := false              if kernel_wanted_to_preempt_me
        if kernel_wanted_to_preempt_me              yield
            yield
        do_not_preempt_me := true
```

To avoid abuse, the kernel is free to ignore the do_not_preempt_me flag if it stays set for too long. It can also deduct any extra time granted a thread from the beginning of its subsequent quantum. Other groups have proposed related mechanisms that can likewise be used to avoid [Marsh et al., 1991] or recover from [Anderson et al., 1992] inopportune preemption. Solaris, in particular [Dice, 2011], provides a schedctl mechanism closely related to that of Edler et al.

The code shown for test_and_set above can easily be adapted to many other locks, with features including backoff, locality awareness, timeout, double-checked or asymmetric locking, and adaptive spin-then-wait. Fair queueing is harder to accommodate. In a ticket, MCS, or CLH lock, one must consider the possibility of preemption not only while holding a lock, but also while waiting in line. So if several threads are waiting, preemption of any one may end up creating a convoy.

To avoid passing a lock to a thread that has been preempted while waiting in line, Kontothanassis et al. [1997] proposed extensions to the kernel interface in the spirit of Edler et al. Specifically, they provide additional values for the do_not_preempt_me flag, and make it visible to other threads. These changes allow one thread to pass a lock to another, and to make the other nonpreemptable, atomically. In a different vein, He et al. [2005] describe a family of queue-based locks in which a lock-releasing thread can estimate (with high confidence) whether the next thread in line has been preempted, and if so dynamically remove it from the queue. The key to these locks is for each spinning thread to periodically write the wall-clock time into its lock queue node. If a thread discovers that the difference between the current time and the time in its successor's queue node exceeds some appropriate threshold, it assumes that the successor is preempted. A thread whose node has been removed from the queue will try again the next time it has a chance to run.

7.5.3 RESOURCE MINIMIZATION

In addition to avoiding unnecessary crossings in and out of kernel mode, the futexes of Linux and the lwp_park-based mutexes of Solaris also eliminate the need to allocate a kernel-level condition queue for every lock. Lists of blocked threads are kept in user-level memory; within the kernel, a waiting thread is simply descheduled, with a note in its context block that indicates what it is waiting for. Since kernel address space is often a scarce resource (especially for preallocated structures), this migration of information into user space may significantly increase the maximum number of locks a system can support.

A similar observation regarding kernel resources was made by the designers of the NT kernel, the foundation for Microsoft OS releases starting with Windows 2000. Compared to Unix variants, the Windows API includes a much larger number of standard library routines. These in turn declare a very large number of internal locks (Windows mutex objects), most of which are never used in a typical program. To economize on kernel resources, NT delays allocating a kernel queue for a given lock until some thread actually tries to acquire the lock. Kernel space is still required for every active lock, and kernel-mode crossings occur on every acquire and release, but space is never wasted on locks that are not used.

Unfortunately, delayed allocation of kernel queues raises the possibility of a run-time exception if space is not available. This possibility was a source of considerable complexity and brittleness in Windows 2000. To eliminate the run-time exception, the designers of Windows XP introduced the notion of *keyed events* [Duffy, 2006], which allow logically distinct conditions to share a single kernel-level queue. Every call to wait or set (signal) must specify both an event and a 32-bit key. Every thread waiting in the queue is then tagged with the key it provided, and set will only awaken a thread with a matching key. Under most circumstances, the kernel allocates a new keyed event for every active lock. If it runs out of memory, however, it falls back to a preexisting (per-address-space) keyed event, with a new lock-specific key. In Windows XP, which used a linked list for the per-address-space queue, performance could sometimes be a problem. Windows Vista replaced the list with a hash table for fast key-based lookups. It also introduced a new family of synchronization objects, including a "slim reader-writer lock," or SRWL. Like futexes and lwp_park-based mutexes, these objects maintain state in user-level memory, and avoid kernel-mode crossings whenever possible. When a thread *does* need to block, it always employs the per-address-space queue.

CHAPTER 8

Nonblocking Algorithms

When devising a concurrent data structure, we typically want to arrange for methods to be atomic—most often linearizable (Section 3.1.2). Most concurrent algorithms achieve atomicity by means of mutual exclusion, implemented using locks. Locks are *blocking*, however, in the formal sense of the word: whether implemented by spinning or rescheduling, they admit system states in which a thread is unable to make progress without the cooperation of one or more peers. This in turn leads to the problems of inopportune preemption and convoys, discussed in Section 7.5.2. Locks—coarse-grain locks in particular—are also typically conservative: in the course of precluding unacceptable thread interleavings, they tend to preclude many acceptable interleavings as well.

We have had several occasions in earlier chapters to refer to *nonblocking* algorithms, in which there is never a reachable state of the system in which some thread is unable to make forward progress. In effect, nonblocking algorithms arrange for every possible interleaving of thread executions to be acceptable. They are thus immune to inopportune preemption. For certain data structures (counters, stacks, queues, linked lists, hash tables—even skip lists) they can also outperform lock-based alternatives even in the absence of preemption or contention.

The literature on nonblocking algorithms is enormous and continually growing. Rather than attempt a comprehensive survey here, we will simply introduce a few of the most widely used nonblocking data structures, and use them to illustrate a few important concepts and techniques. A more extensive and tutorial survey can be found in the text of Herlihy and Shavit [2008]. Håkan Sundell's Ph.D. thesis [2004] and the survey of Moir and Shavit [2005] are also excellent sources of background information. Before proceeding here, readers may wish to refer back to the discussion of *liveness* in Section 3.2.

8.1 SINGLE-LOCATION STRUCTURES

The simplest nonblocking algorithms use the CAS and LL/SC-based fetch_and_Φ constructions of Section 2.3 to implement methods that update a single-word object. An atomic counter (accumulator) object, for example, might be implemented as shown in Figure 8.1. Reads (get) and writes (set) can use ordinary loads and stores, though the stores must be write atomic to avoid causality loops. Updates similarly require that fetch_and_Φ instructions be write atomic. Note that in contrast to the lock algorithms of Chapter 4, we have not employed any fences or other synchronizing instructions to order the operations of our object with respect to preceding

```
class counter                          int counter.increase(int v):
    int c                                  int old, new
                                           repeat
int counter.get():                             old := c
    return c                                   new := old + v
                                           until CAS(&c, old, new)
void counter.set(int v):                   return old
    c := v
```

Figure 8.1: A single-word atomic counter, implemented with CAS. If updates to the counter are to be seen in consistent order by all threads, the store in set and the CAS in increase must both be write atomic.

```
class stack                            node* stack.pop():
    ⟨node*, int⟩ top                       repeat
                                               ⟨o, c⟩ := top
void stack.push(node* n):                      if o = null return null
    repeat                                     n := o→next
        ⟨o, c⟩ := top                      until CAS(&top, ⟨o, c⟩, ⟨n, c+1⟩)
        n→next := o                        return o
    until CAS(&top, ⟨o, c⟩, ⟨n, c⟩)
```

Figure 8.2: The lock-free "Treiber stack," with a counted top-of-stack pointer to solve the ABA problem (reprised from Figure 2.7). It suffices to modify the count in pop only; if CAS is available in multiple widths, it may be applied to only the pointer in push.

or following code in the calling thread. If such ordering is required, the programmer will need to provide it.

Slightly more complicated than a single-word counter is the lock-free stack of Section 2.3.1, originally published by Treiber [1986] for the IBM System/370, and very widely used today. Code for this stack is repeated here as Figure 8.2. As discussed in the earlier section, a sequence count has been embedded in the top-of-stack pointer to avoid the ABA problem. Without this count (or some other ABA solution [Jayanti and Petrovic, 2003, Michael, 2004a]), the stack would not function correctly.

If our stack had additional, read-only methods (e.g., an is_empty predicate), then the CASes that modify top would need to be write atomic. Similar observations will apply to other nonblocking data structures, later in this chapter. In an algorithm based on mutual exclusion or reader-writer locks, linearizability is trivially ensured by the order of updates to the lock. With seqlocks or RCU, as noted in Sections 6.2 and 6.3, write atomicity is needed to ensure that readers see updates in consistent order. In a similar way, any updates in a nonblocking data structure that might otherwise appear inconsistent to other threads (whether in read-only operations or in portions of more general operations) will need to be write atomic.

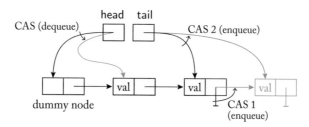

Figure 8.3: Operation of the M&S queue. After appropriate preparation ("snapshotting"), dequeue reads a value from the second node in the list, and updates head with a single CAS to remove the old dummy node. In enqueue, two CASes are required: one to update the next pointer in the previous final node; the other to update tail.

8.2   THE MICHAEL AND SCOTT (M&S) QUEUE

Queues are substantially more difficult than stacks to implement in a nonblocking fashion. The most commonly used solution is due to Michael and Scott [1998, 1996]. It appears in most operating systems, and in a wide variety of run-time systems. The authors report that it rivals or exceeds the performance of lock-based alternatives on all machines they tested. Like the Treiber stack, the M&S queue is lock free but not wait free: it precludes livelock, but admits the possibility of starvation. A single enqueue and a single dequeue can proceed concurrently on a nonempty queue, without interfering with one another. If multiple enqueues or multiple dequeues are active concurrently, one of each is guaranteed to complete in a bounded number of steps; the others back out and retry.

An illustration of the M&S queue appears in Figure 8.3. Code appears in Figure 8.4. Where the Treiber stack expected a node to be passed to push and returned from pop, we have written enqueue to expect (and dequeue to return) a simple value. We assume the existence of a distinguished error value ⊥ that will be returned on any attempt to dequeue an item from an empty queue. One could, of course, return a separate status code or throw an exception instead.

To understand the behavior of the queue, it helps to consider the individual instructions—the *linearization points* (Section 3.1.2)—at which operations can be considered to take effect. If we choose these properly, then whenever the linearization point of operation A precedes the linearization point of operation B, we will know that operation A, as a whole, linearizes before operation B.

Dummy Nodes.   To avoid special cases found in prior algorithms, the M&S queue always keeps a "dummy" node at the head of the queue. The first real item is the one in the node, if any, that *follows* the dummy node. As each item is dequeued, the old dummy node is reclaimed, and the node in which the dequeued item was located becomes the new dummy node.

```
type ptr = ⟨node* p, int c⟩                          // counted pointer
type node
      value val
      ptr next
class queue
      ptr head
      ptr tail

void queue.init()
      node* n := new node(⊥, null)                   // initial dummy node
      head.p := tail.p := n

void queue.enqueue(value v):
      node* w := new node(v, null);  fence(W‖W)       // allocate node for new value
      ptr t, n
      loop
          t := tail.load()                           // counted pointers
          n := t.p→next.load()
          if t = tail.load()                         // are t and n consistent?
              if n.p = null                          // was tail pointing to the last node?
                  if CAS(&t.p→next, n, ⟨w, n.c+1⟩)    // try to add w at end of list
                      break                          // success; exit loop
              else                                   // tail was not pointing to the last node
                  (void) CAS(&tail, t, ⟨n.p, t.c+1⟩) // try to swing tail to next node
      (void) CAS(&tail, t, ⟨w, t.c+1⟩)               // try to swing tail to inserted node

value queue.dequeue():
      ptr h, t, n
      loop
          h := head.load()                           // counted pointers
          t := tail.load()
          n := h.p→next.load()
          value rtn
          if h = head.load()                         // are h, t, and n consistent?
              if h.p = t.p                           // is queue empty or tail falling behind?
                  if n.p = null return ⊥             // empty; return failure
                  (void) CAS(&tail, t, ⟨n.p, t.c+1⟩) // tail is falling behind; try to update
              else                                   // no need to deal with tail
                  // read value before CAS; otherwise another dequeue might free n
                  rtn := n.p→val.load()
                  if CAS(&head, h, ⟨n.p, h.c+1⟩)      // try to swing head to next node
                      break                          // success; exit loop
      fence(W‖W)                                      // link node out before deleting!
      free_for_reuse(h.p)                             // type-preserving
      return rtn                                      // queue was nonempty; return success
```

Figure 8.4: The M&S lock-free queue, with counted pointers to solve the ABA problem.

Swinging the Tail.   A successful dequeue operation linearizes straightforwardly at the CAS that moves the head pointer. An unsuccessful dequeue (one that returns $\perp$) linearizes at the load of n in the last iteration of the loop. An enqueue operation is somewhat trickier: it requires a *pair* of updates: one to make the next field of the most recent previously enqueued node point to the new node, and one to make the tail pointer point to it. The update to the next pointer constitutes the linearization point; the update of the tail is cleanup, and can be performed by any thread.

Consistent Snapshots.   Prior to its linearization point, enqueue must read both the tail pointer and, subsequently, tail.p→next. To ensure that these reads are mutually consistent—that no other enqueue linearizes in-between—the code re-reads tail after reading tail.p→next. Since every change to tail increments tail.c, we can be sure (absent rollover) that there was a moment in time at which both locations held the values read, simultaneously. A subsequent CAS of tail.p→next will be attempted only if tail.p→next.p is null; given the use of counted pointers, the CAS will be successful only if tail.p→next has not changed since it was read. Since a node is never removed from the list unless it has a non-null successor (which will become the new dummy node), we can be sure that our new node will be linked into the queue only as the successor to an end-of-list node that has not been re-used since it was pointed at by tail.

   The dequeue operation, for its part, re-reads head after reading tail and head.p→next. Since an enqueue operation can move tail farther away from head, but never closer, we can be sure, if the values read for head and tail are not equal, that the queue cannot become empty until head changes: a successful CAS that expects the original value of head cannot cause head to move past tail. Similarly, since a node is never removed from the queue without changing head, a successful CAS that expects the original value of head can be sure that the previously read value of head.p→next is the right value to install as the new head.

Memory Management.   If a thread in dequeue is delayed (perhaps because of preemption) immediately before its read of n.p→val, it may end up performing the read long after the node has been removed from the queue and reused by another thread. More significantly, if a thread in enqueue is delayed immediately before its initial CAS, it may attempt to modify the next pointer of a node that has been removed from the queue and reused. To ensure the safety of such delayed accesses, it is essential that the space occupied by a removed node never be returned to the operating system (allowing an access to trigger a segmentation fault), and that the word occupied by the node's next pointer never be reused for a purpose that might accidentally lead to the same bit pattern used for the counted pointer. In a pattern common to many nonblocking pointer-based algorithms, Michael and Scott employ a *type-preserving* allocator, as described in the box on page 26.

   In recent years, several techniques have been proposed to improve the performance of nonblocking queues. Ladan-Mozes and Shavit [2008] effect an enqueue with a single CAS by using an MCS-style list in which the operation linearizes at the update of the tail pointer, and the forward link is subsequently created with an ordinary store. If a dequeue-ing thread finds that

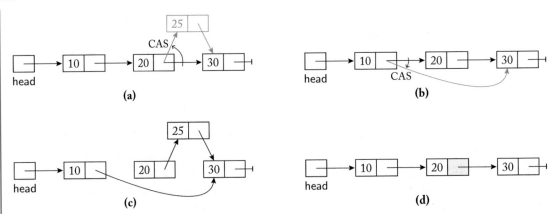

Figure 8.5: Atomic update of a singly linked list. Naive insertion (a) and deletion (b), if executed concurrently, can leave the list in an inconsistent state (c). The H&M list therefore performs a two-step deletion that first marks the next pointer of the to-be-deleted node (shown here with shading) (d), thereby preventing conflicting insertion.

the next node has not yet been "linked in" (as may happen if a thread is delayed), it must traverse the queue from the tail to fix the broken connection. Hendler et al. [2010b] use *flat combining* (Section 5.4) to improve locality in high-contention workloads by arranging for multiple pending operations to be performed by a single thread. Morrison and Afek [2013] observe that both the overhead of memory management and the contention caused by failed-and-repeating CAS operations can be dramatically reduced by storing multiple data items in each element of the queue, and using fetch_and_increment to insert and delete them.

8.3   HARRIS AND MICHAEL (H&M) LISTS

Singly-linked lists are among the most ubiquitous data structures in computer science. Sorted linked lists, in particular, are frequently used as a simple implementation of a set, and may also be used as a component of more complex data structures (e.g., hash tables and skip lists). Sundell and Tsigas [2008b] have shown that even doubly linked lists can be manipulated in a lock-free fashion, though the code to do so is quite complex. We focus here on the singly linked case.

It is tempting to expect that both insertion of a node into a list and removal of a node from a list would be trivial to implement in a nonblocking fashion using CAS, since in each case only a single pointer of the existing list needs to change. In Figure 8.5(a), insertion of a node containing the value 25 requires only that we swing the next pointer of the node containing 20. In Figure 8.5(b), deletion of the node containing 20 requires only that we swing the next pointer of the node containing 10.

The problem with this expectation is that insertion after the node containing 20 is correct only if that node remains a member of the list. If the deletion in 8.5(b) happens immediately before the insertion in 8.5(a), we shall be left with the configuration of 8.5(c), in which the node containing 25 has inadvertently been lost.

The solution to this problem, first used in an early queue algorithm by Prakash et al. [1994] and then adapted to the general linked-list case by Harris [2001], is to perform deletion in two steps. The first step, shown in Figure 8.5(d), marks the to-be-deleted node by flipping a bit in its own next pointer (this is suggested in the figure by shading the box containing the pointer). A second, separate step (not shown in the figure) updates the next pointer of the node containing 10 to actually remove the deleted node from the list. In between these two steps, the list remains intact, and can safely be traversed, but lookup operations will decline to "see" the deleted node, and insert operations will decline to update its marked next pointer. Moreover, because the state of the list remains well defined, and the second step can be effected with a single CAS, any thread that comes upon a deleted node can safely remove it, without waiting for help from any other thread. Assuming the desired key was in the list, deletion linearizes at the initial, pointer-marking CAS; assuming the desired key was absent, insertion linearizes at the single CAS that adds its node to the list. (Linearization points differ for lookups and for failed deletions and insertions; see below.)

As in the M&S queue, the ABA problem becomes an issue if nodes can be reused. Harris's algorithm assumes the existence of general-purpose garbage collection (e.g., based on reference counts) that will refrain from reclaiming nodes until all existing references have disappeared. Michael [2002] refined the algorithm to accommodate memory management based either on counted pointers and a type-preserving allocator (as in the M&S queue) or on hazard pointers [Michael, 2004b].

The counted pointer version of the list, adapted from Michael's paper and augmented with appropriate synchronizing instructions, appears in Figures 8.6 and 8.7. While it requires a double-width CAS to update a pointer and count simultaneously, it has the advantage of easy adaptation to applications in which a node must be moved from one list to another, without waiting for stale references to expire.

The heart of the Harris & Michael algorithm is the search routine, shown in Figure 8.6, and called in turn by insert, delete, and lookup (Figure 8.7). For the sake of notational convenience, search is designed to "return" three values—PREVp, CURR, and NEXT as thread-local variables. CURR is a counted-pointer reference to the first node, if any, with a key greater than or equal to the searched-for value. PREVp is a conventional pointer to the counted pointer in the list that refers to CURR. NEXT is a counted-pointer reference to the node, if any, that follows CURR. The various cases that arise near the beginning and end of the list are illustrated in Figure 8.8.

In Section 3.1.2 we mentioned the existence of algorithms in which a run-time check allows a method to determine that it linearized at some previous instruction. The H&M list is such an algorithm. As noted above, successful insertions and deletions linearize at CAS operations

```
type ptr = ⟨bool d, node* p, int c⟩                    // markable counted pointer
type node
    value val
    ptr next
class list
    ptr head

// thread-private variables, changed by search:
ptr* PREVp
ptr CURR, NEXT

// Find first node N (or null) where val is ≥ v. On return, CURR refers to N (or is null);
// PREVp points at ptr that referred to N; NEXT caches N's next ptr (if any).
private bool list.search(value v):
    loop
        PREVp := &head
        CURR := *PREVp.load()
        loop
            if CURR.p = null                           // v is bigger than anything in list
                return false
            NEXT := CURR.p→next.load()
            value cv := CURR.p→val.load()
            if *PREVp.load() ≠ ⟨false, CURR.p, CURR.c⟩
                continue outer loop                    // list has been changed; start over
            if NEXT.d
                // node has been marked deleted; attempt to link it out
                if CAS(PREVp, ⟨false, CURR.p, CURR.c⟩, ⟨false, NEXT.p, CURR.c+1⟩)
                    fence(W‖W)                          // link node out before deleting!
                    free_for_reuse(CURR.p)              // type-preserving
                    NEXT.c := CURR.c+1
                else continue outer loop                // list has been changed; start over
            else
                if cv ≥ v return cv = v
                PREVp := &CURR.p→next
            CURR := NEXT
```

Figure 8.6: The H&M lock-free list, as presented by Michael [2002] (definitions and internal search routine), with counted pointers to solve the ABA problem. Synchronizing instructions have been added to the original. CAS instructions are assumed to be write atomic.

that occur after the call to search. Unsuccessful insertions (attempts to insert an already present key), unsuccessful deletions (attempts to remove an already missing key), and all calls to lookup linearize within the search routine. If the list is empty, the linearization point is the load of CURR from *PREVp, immediately before the inner loop; if the list is non-empty, it is the last dynamic load of NEXT from CURR.p→next, at the third line of the last iteration of the inner loop. In all these intra-search cases, we don't know that the method has linearized until we inspect the loaded value.

```
bool list.insert(value v):
    if search(v) return false
    node* n := new node(v, ⟨false, CURR.p, 0⟩)
    loop
        // note that CAS is ordered after initialization/update of node
        if CAS(PREVp, ⟨false, CURR.p, CURR.c⟩, ⟨false, n, CURR.c+1⟩, W‖)
            return true
        if search(v)
            free(n)              // node has never been seen by others
            return false
        n→next := ⟨false, CURR.p, 0⟩
bool list.delete(value v):
    loop
        if ¬search(v) return false
        // attempt to mark node as deleted:
        if ¬CAS(&CURR.p→next, ⟨false, NEXT.p, NEXT.c⟩, ⟨true, NEXT.p, NEXT.c+1⟩)
            continue                          // list has been changed; start over
        // attempt to link node out of list:
        if CAS(PREVp, ⟨false, CURR.p, CURR.c⟩, ⟨false, NEXT.p, CURR.c+1⟩)
            fence(W‖W)                        // link node out before deleting!
            free_for_reuse(CURR.p)            // type-preserving
        else (void) search(v)    // list has been changed; re-scan and clean up deleted node(s)
        return true
bool list.lookup(value v):
    return search(v)
```

Figure 8.7: The H&M lock-free list (externally visible methods). Note in Figure 8.6 that PREVp, CURR, and NEXT are thread-private variables changed by list.search.

8.4   HASH TABLES

In his paper on nonblocking lists, Michael [2002] presented a straightforward extension of his code to yield a nonblocking hash table with external chaining and a fixed number of buckets. Each bucket of the table is the head pointer of a nonblocking list; lookup, insert, and delete simply apply the identically-named list method to the appropriate bucket.

The problem with this approach is that the size of the table cannot easily change. If we lack a good *a priori* estimate of the number of items that will eventually belong to our set, we need an *extensible* hash table. We can obtain one by protecting the table with a single sequence lock (Section 6.2), which ordinary lookup, insert, and delete methods access as "readers," and which a thread that chooses to enlarge the table acquires as a "writer." This strategy preserves safety but not nonblocking progress: when a resize is in progress, lookup, insert, and delete operations must wait. If we use RCU (Section 6.3) to delay reclamation of the older, smaller table, we can allow lookup operations to proceed in parallel with resizing, but insert and delete will still need to wait.

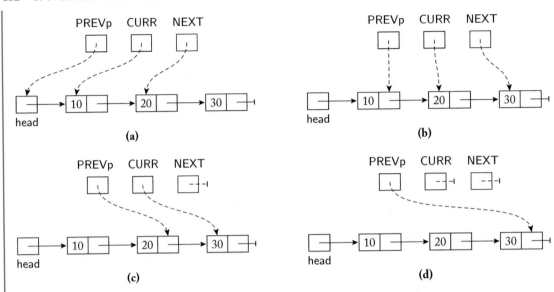

Figure 8.8: Searching within an H&M list. PREVp is a pointer to a markable counted pointer (ptr); CURR and NEXT are ptrs. Diagrams (a), (b), (c), and (d) show the final positions of PREVp, CURR, and NEXT when the searched-for value is ≤ 10, ≤ 20, ≤ 30, and > 30, respectively. The return value of search will be true if and only if the value is found precisely at *CURR.

Ideally, we should like resizing to be a nonblocking operation that allows not only lookup but also insert and delete operations to continue unimpeded. Shalev and Shavit [2006] describe an algorithm that achieves precisely this objective. It is also *incremental*: the costs of a resizing operation are spread over multiple insert, delete, and lookup operations, retaining $O(1)$ expected time for each. The basic idea is illustrated in Figure 8.9. Instead of a separate chain of nodes in each bucket, it maintains a single list of all nodes, sorted by *order number*. Given a hash function h with a range of $0 \ldots 2^n - 1$, we obtain the order number of a node with key k by reversing the n bits of $h(k)$ and then adding an extra least-significant 1 bit.

Fast access into the list of nodes is provided by a collection of 2^j lazily initialized buckets, where j is initialized to some small positive integer value i, and may increase at run time (up to a limit of n) to accommodate increases in the length of the list. Each initialized bucket contains a pointer to a so-called *dummy* node, linked into the list immediately before the data nodes whose top j order number bits, when reversed, give the index of the bucket. To ensure that it appears in the proper location, the dummy node for bucket b is given an order number obtained by reversing the j bits of b, padding on the right with $n - j$ zeros, and adding an extra least-significant 0 bit.

The point of all this bit manipulation is to ensure, when we decide to increment j (and thus double the number of buckets), that all of the old buckets will still point to the right places

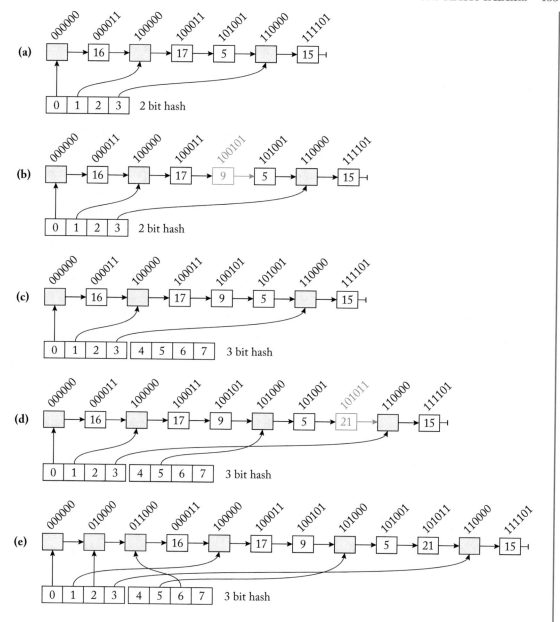

Figure 8.9: The nonblocking, extensible S&S hash table. Dummy nodes are shaded. Data nodes are labeled (for illustration purposes) with the hash of their key; order numbers are shown above. Starting from the configuration shown in (a), we have inserted a data node with hash value 9 (b), doubled the number of buckets (c), inserted a node with hash value 21 (d), and searched for a node with hash value 30 (e).

in the list, and the new buckets, once they are initialized, will point to new dummy nodes *interspersed among* the old ones. We can see this dynamic at work in Figure 8.9. Part (a) shows a table containing 4 elements whose keys are not shown, but whose hash values are 5, 15, 16, and 17. For simplicity of presentation, we have assumed that $n = 5$, so hash values range from 0 to 31. Above each node we have shown the corresponding order number. The node with hash value 5, for example, has order number $(00101_2)^R << 1 + 1 = 101001_2$.

Still in (a), we use the two low-order bits of the hash to index into an array of $2^2 = 4$ buckets. Slots 0, 1, and 3 contain pointers to dummy nodes. All data whose hash values are congruent to b mod 4 are contiguous in the node list, and immediately follow the dummy node for bucket b. Note that bucket 2, which would not have been used in the process of inserting the four initial nodes, is still uninitialized. In (b) we have inserted a new data node with hash value 9. It falls in bucket 1, and is inserted, according to its order number, between the nodes with hash values 17 and 5. In (c) we have incremented j and doubled the number of buckets in use. The buckets themselves will be lazily initialized.

To avoid copying existing buckets (particularly given that their values may change due to lazy initialization), we employ noncontiguous bucket arrays of exponentially increasing size. In a simplification of the scheme of Shalev and Shavit, we access these arrays through a single, second-level directory (not shown). The directory can be replaced with a single CAS. It indicates the current value of j and the locations of $j - 1$ bucket arrays. The first two arrays are of size 2^i (here $i = 2$); the next is of size $2^{(i+1)}$, and so on. Given a key k, we compute $b = h(k)$ mod 2^j and $d = b >> i$. If $d = 0$, b's bucket can be found at directory[0][b mod 2^i]. Otherwise, let m be the index of the most significant 1 bit in d's binary representation; b's bucket can be found at directory[$m + 1$][b mod 2^{m+i}].

In (d) we have inserted a new data node with hash value 21. This requires initialization of bucket $(21$ mod $2^3) = 5$. We identify the "parent" of bucket 5 (namely, bucket 1) by zeroing out the most significant 1 bit in 5's binary representation. Traversing the parent's portion of the node list, we find that 5's dummy node (with order number 101000_2) belongs between the data nodes with hash values 9 and 5. Having inserted this node, we can then insert the data node with hash value 21. Finally, in (e), we search for a node with hash value 30. This requires initialization of bucket $(30$ mod $2^3) = 6$, which recursively requires initialization of bucket 6's parent—namely bucket 2. Shalev and Shavit prove that the entire algorithm is correct and nonblocking, and that given reasonable assumptions about the hash function h, the amortized cost of insert, delete, and lookup operations will be constant.

8.5 SKIP LISTS

Implementations of sets and dictionaries with $O(\log n)$ search time are commonly based on trees, but these are notoriously difficult to parallelize, particularly in a nonblocking fashion. A skip list [Pugh, 1990] is an alternative structure that uses randomness rather than rebalancing to achieve expected logarithmic time. While nonblocking skip lists are still quite complex—perhaps the most complex nonblocking structures currently in widespread use—they are substantially simpler than nonblocking trees. (For a promising example of the latter, see recent work by Howley and Jones [2012].)

All the nodes in a skip list appear on a single, sorted, "level 0" list. With probability p (typically $1/2$), each node also appears on a level 1 list. In general, a node on a level i list appears on a level $i + 1$ list with probability p. All nodes with a given key are linked together in a "tower." Searching begins by traversing the high level lists, which skip many nodes, and then moving down towers to "zero in" with lower-level lists.

Fraser [2003] is credited with the first nonblocking skip list implementation. Variants of this implementation (the ConcurrentSkipListSet and ConcurrentSkipListMap) were developed by Doug Lea, and appear in the standard concurrency library for Java. Herlihy and Shavit [2008, Sec. 14.4] describe a further refinement due to Herlihy, Lev, and Shavit. Additional, more recent variants can be found in the literature. All are based on lock-free lists.

8.6   DOUBLE-ENDED QUEUES

Unlike a stack, which permits insertions and deletions at one end of a list, and a queue, which permits insertions at one end and deletions at the other, a double-ended queue, or *deque* (pronounced "deck" or "deek") permits insertions and deletions at both ends (but still not in the middle). In comparison to stacks and queues, deques have fewer practical uses. The most compelling, perhaps, is the elegant $O(n)$ convex hull algorithm of Melkman [1987]. The most familiar is probably the *history* or *undo* list of an interactive application: new operations are pushed onto the head of the list, undone operations are popped off the head, and old operations are dropped off the tail as the list continues to grow (there are, however, no insertions at the tail).

For nonblocking concurrent programming, deques have long been a subject of intrinsic intellectual interest, because they are more complex than stacks and queues, but still simpler than structures like search trees. The standard CAS-based lock-free deque is due to Michael [2003]; we describe it in Section 8.6.1 below. We then consider, in Section 8.6.2, an algorithm due to Herlihy et al. [2003a] that achieves a significant reduction in complexity by using obstruction freedom rather than lock freedom as its liveness criterion. Michael's queues employ an unbounded, doubly linked list; those of Herlihy et al. employ a circular array. Other algorithms can be found in the literature; in particular, Sundell and Tsigas [2008b] use their lock-free doubly linked lists to construct an unbounded nonblocking dequeue in which operations on the head and tail can proceed in parallel.

In addition to uses inherited from sequential programming, concurrent deques have a compelling application of their own: the management of tasks in a *work-stealing scheduler*. We consider this application in Section 8.6.3.

8.6.1   UNBOUNDED LOCK-FREE DEQUES

The lock-free deque of Michael [2003] uses a single, double-width, CAS-able memory location (the "anchor") to hold the head and tail pointers of the list, together with a 2-bit status flag that can take on any of three possible values: STABLE, LPUSH, and RPUSH. For ABA-safe memory allocation, the algorithm can be augmented with hazard pointers [Michael, 2004b]. Alternatively, it can be modified to rely on counted pointers, but to fit two of these plus the status flag in a single CAS-able anchor—even one of double width—the "pointers" must be indices into a bounded-size pool of nodes. If this is unacceptable, double-wide LL / SC can be emulated with an extra level of indirection [Michael, 2004a].

Operations on the deque are illustrated in Figure 8.10. At any given point in time, the structure will be in one of seven functionally distinct states. Blue arrows in the figure indicate state transitions effected by push_left, pop_left, push_right, and pop_right operations (arrows labeled simply "push" or "pop" cover both left and right cases).

Three states are STABLE, as indicated by their status flag: they require no cleanup to complete an operation. In S_0 the dequeue is empty—the head and tail pointers are null. In S_1 there is a single node, referred to by both head and tail. In S_{2+} there are two or more nodes, linked together with left and right pointers.

Four states—those with status flags LPUSH and RPUSH—are transitory: their contents are unambiguous, but they require cleanup before a new operation can begin. To ensure nonblocking progress, the cleanup can be performed by any thread. In a push_right from state S_2, for example, an initial CAS changes the status flag of the anchor from STABLE to RPUSH and simultaneously updates the tail pointer to refer to a newly allocated node. (This node has previously been initialized to contain the to-be-inserted value and a left pointer that refers to the previous tail node.) These changes to the anchor move the deque to the *incoherent* state R_i, in which the right pointer of the second-to-rightmost node is incorrect. A second CAS fixes this pointer, moving the deque to the *coherent* state R_c; a final CAS updates the status flag, returning the deque to state S_{2+}.

The actual code for the deque is quite complex. Various operations can interfere with one another, but Michael proves that an operation fails and starts over only when some other operation has made progress.

8.6.2   OBSTRUCTION-FREE BOUNDED DEQUES

The principal shortcoming of Michael's deque is the need to colocate the head and tail pointers. This not only raises issues of how many bits will fit in a CAS-able word: it also implies that all operations on the deque interfere with one another. While nonblocking, they must serialize on their access to the anchor word. Ideally, we should like to arrange for operations on opposite ends of a nonempty deque to proceed entirely in parallel.

At about the same time that Michael was devising his algorithm, a group at Sun (now Oracle) Labs in Boston was developing the notion of *obstruction freedom* [Herlihy et al., 2003a],

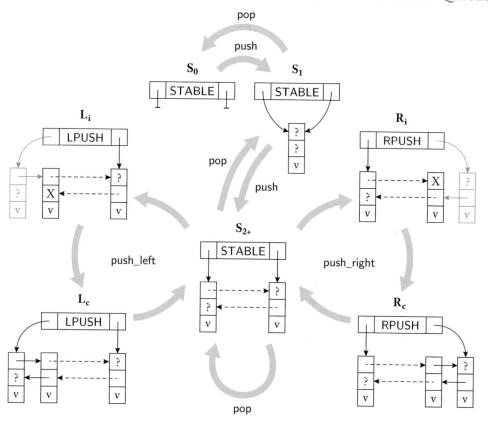

Figure 8.10: Operation of a lock-free deque (figure adapted from Michael [2003]). Blue arrows indicate state transitions. In each state, the anchor word is shown at top, comprising the head pointer, status flag, and tail pointer. Nodes in the queue (oriented vertically) contain right and left pointers and a value ('v'). Interior nodes are elided in the figure, as suggested by dashed arrows. A left or right push from state S_2 is a three-step process. Nodes in the process of being inserted are shown in gray. Question marks indicate immaterial values, which will not be inspected. An 'X' indicates a temporarily incorrect (*incoherent*) pointer.

which we introduced in Section 3.2.1. Where a lock-free algorithm (such as Michael's) guarantees that some thread will make progress within a bounded number of steps (of any thread), an obstruction-free algorithm guarantees only that a thread that runs *by itself* (without interference from other threads) will always make progress, regardless of the starting state. In effect, Herlihy et al. argued that since a lock-free algorithm already requires some sort of contention management mechanism (separate from the main algorithm) to avoid the possibility of starvation, one might as well ask that mechanism to address the possibility of livelock as well, thereby separating issues

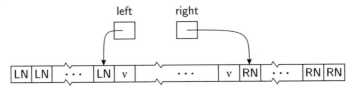

Figure 8.11: The HLM obstruction-free deque. Each 'v' represents a data value. 'LN' is a *left null* value; 'RN' is a *right null* value. The left and right (head and tail) pointers are hints; they point at or near the rightmost LN and leftmost RN slots in the array.

of safety and liveness entirely. By doing so, the authors argue, one may be able to simplify the main algorithm considerably. Double-ended queues provide an illustrative example. Nonblocking versions of transactional memory (Chapter 9) provide another.

Michael's lock-free deque employs a linked list whose length is limited only by the range of pointers that will fit in the anchor word. By contrast, the deque of Herlihy et al. employs a fixed-length circular array. It is most easily understood by first considering a noncircular version, illustrated in Figure 8.11. At any given time, reading from left to right, the array will contain one or more LN ("left null") values, followed by zero or more data values, followed by one or more RN ("right null") values. To perform a right push, one must replace the leftmost RN with a data value; to perform a right pop, one must read the rightmost data value and replace it with an RN. The left-hand cases are symmetric. To find the leftmost RN, one can start at any entry of the array: if it's an RN, scan left to find the last RN; it it's an LN or data value, scan right to find the first RN. To reduce the time consumed, it is helpful to know approximately where to start looking, but the indication need not be exact.

Given these observations, the only two really tricky parts of the algorithm are, first, how to make sure that every operation maintains the LN...v...RN structural invariant, and, second, how to join the ends of the array to make it circular.

The first challenge is addressed by adding a count to every element of the array, and then arranging for every operation to modify, in an appropriate order, a *pair* of consecutive elements. A right push operation, for example, identifies the index, k, of the leftmost RN value. If k is the rightmost slot in the array, the operation returns a "deque is full" error message. Otherwise, it performs a pair of CASes. The first increments the count in element $k - 1$; the second replaces element k with a new data value and an incremented count. A right pop operation goes the other way: it identifies the index, j, of the rightmost data value (if any). It then performs its own pair of CASes. The first increments the count in element $j + 1$; the second replaces element j with RN and an incremented count. Left-hand operations are symmetric.

The key to linearizability is the observation that only the second CAS of a pair changes the actual content of the deque; the first ensures that any conflict with a concurrent operation will be noticed. Since we read both locations ($k - 1$ and k, or $j + 1$ and j) before attempting a

CAS on either, if both CASes succeed, no other operation modified either location in-between. If the first CAS fails, no change has been made. If the second CAS fails, no *substantive* change has been made. In either case, the operation can simply start over. Updates to the global left and right pointers constitute cleanup. Because the pointers are just hints, atomicity with the rest of the operation is not required. Moreover, updates to left and right need not interfere with one another.

It is easy to see that the algorithm is obstruction free: an operation that observes an unchanging array will always complete in a bounded number of steps. It is also easy to see that the algorithm is not lock free: if a right push and a right pop occur at just the right time, each can, in principle, succeed at its first CAS, fail at the second, and start over again, indefinitely. A right push and a left push on an empty deque can encounter a similar cycle. In practice, randomized backoff can be expected to resolve such conflicts quickly and efficiently.

To make the deque circular (as indeed it must be if pushes and pops at the two ends are not precisely balanced), Herlihy et al. introduce one new *dummy null* (DN) value. The structural invariant is then modified to allow the empty portion of the circular array to contain, in order, zero or more RN values, zero or one DN values, and zero or more LN values. At all times, however, there must be null values of at least two different kinds—at least one RN or DN, and at least one DN or LN.

A right push that finds only one RN value in the array must change the adjacent DN value, if any, into an RN first. If there is no adjacent DN, the operation must change the leftmost LN, if any, into a DN first. In all cases, changes are made with a pair of CASes, the first of which increments a count and the second of which is substantive.

8.6.3   WORK-STEALING QUEUES

In parallel programming languages and libraries, it is commonplace to distinguish between *threads*, which are often provided by the operating system in rough proportion to the degree of hardware parallelism, and *tasks*, which are logically distinct units of work to be accomplished. Worker threads then execute logical tasks under the direction of a user-level scheduler.

Originally developed for the Cilk programming language [Blumofe et al., 1995, Frigo et al., 1998], *work stealing* [Blumofe and Leiserson, 1994] has become the scheduling discipline of choice for tasks. To minimize contention and maximize locality, the scheduler maintains a separate pool of tasks for each worker thread. Upon creation, tasks are inserted into the pool of the creating thread, and when the current task completes, a new one is chosen from this pool. Only when the local pool is empty does a thread go looking for work elsewhere. Strategies here differ: in the simplest case, the thread chooses a random peer and attempts to remove a task from that peer's pool.

Because tasks may be quite small, it is important that insertions and deletions from the local pool be very fast. Toward that end, Arora, Blumofe, and Plaxton [1998] developed a special-purpose deque that is carefully optimized for the work stealing case. It assumes that push_right and pop_right operations are performed by a single thread (the local one), and thus need not be synchronized with one another. It also assumes that push_left operations never occur. Synchronization is required only among pop_left operations and, when a deque is nearly empty, between those operations and any concurrent push_right or pop_right operations.

The ABP algorithm, as it is sometimes known, is both simple and clever. It has very low constant-time overhead in the common case, and is very widely used. It does, however, have two important limitations. First, it uses a bounded array, which limits the number of tasks that can be pushed into the deque at any given time. Second, when tasks are "stolen" via pop_left operations, the space cannot be reclaimed until the deque is empty; that is, the left end index resets to zero only when the local thread "bumps into it" in the course of a pop_right operation. Chase and Lev [2005] present an ABP extension that addresses both limitations: it treats the array as circular, avoiding the reset problem, and it allows the array to be resized on overflow, much as in an extensible hash table. In an attempt to improve performance when workloads are unevenly distributed, Hendler and Shavit [2002] describe another ABP extension in which a thread can, in a single operation, steal up to half the elements from a peer's deque. Many additional extensions and alternatives can be found in the literature; work stealing remains an active topic of research.

8.7   DUAL DATA STRUCTURES

Under even the weakest notion of nonblocking progress, a thread must be able to complete an operation in a bounded number of steps in the absence of activity in other threads. This implies that the operation must be *total*; that is, it must be valid and well defined given any consistent state of the data structure, with no nontrivial preconditions. In defining operations on containers (stacks, queues, etc.), we have assumed that a remove (pop, dequeue) operation on an empty container returns a special null or ⊥ value to indicate its failure.

But this is often not what we want. In many algorithms, a thread that encounters an empty container (or a full bounded container, or an account with insufficient funds, or …) really needs to wait (using condition synchronization) for activity in some other thread to make the needed precondition true.

How do we reconcile the need for condition synchronization with the desire for nonblocking progress? The obvious option is to spin:

```
datum v
repeat
    v := my_container.remove()
until v ≠ ⊥
```

In addition to wasting cycles and increasing contention (as spinning often does), this option has the additional disadvantage that when a new datum is finally inserted into an empty container, the thread that gets to remove it will be determined, more or less accidentally, by the underlying scheduler, rather than by the code of the data structure's methods. To bring the choice under data structure control and, optionally, avoid the use of spinning, Scherer and Scott [2004] developed the notion of nonblocking *dual* data structures.

In addition to data, a dual data structure may also hold *reservations*. When an operation discovers that a precondition does not hold, it inserts a reservation, with the expectation that some subsequent operation (in another thread) will notify it when the precondition holds. The authors describe a formal framework in which both the initial insertion of a reservation and the eventual successful completion of an operation (once the precondition holds) are nonblocking and linearizable, and any intermediate activity (spinning or blocking) results in only a constant number of remote memory operations, and thus can be considered harmless.

As examples, Scherer and Scott present nonblocking dual versions of the Treiber stack (Section 8.1) and the M&S queue (Section 8.2). In both, a remove operation must determine whether to remove a datum or insert a reservation; an insert operation must determine whether to insert a datum or *fulfill* a reservation. The challenge is to make this decision and then carry it out atomically, as a single linearizable operation. Among other things, we must ensure that if operation x satisfies the precondition on which thread t is waiting, then once x has linearized (and chosen t as its successor), t must complete its operation within a bounded number of (its own) time steps, with no other linearizations in between.

In the nonblocking *dualqueue*, atomicity requires a small extension to the consistent snapshot mechanism and a convention that tags each next pointer with a bit to indicate whether the next node in the queue contains a datum or a reservation. (The only tag that is ever inspected is the one in the next pointer of the dummy node: one can prove that at any given time the queue will consist entirely of data nodes or entirely of reservations.) Fulfillment of reservations is straightforward: if a waiting thread spins on a field in the reservation node, we can use a CAS to change that field from ⊥ to the fulfilling datum before removing the node from the queue. (Alternatively, we could signal a single-use condition variable on which the waiting thread was blocked.)

In the nonblocking *dualstack*, next pointers are also tagged, but the lack of a dummy node, and the fact that insertions and deletions occur at the same end of the list, introduces an extra complication. To ensure nonblocking progress, we must fulfill a request before popping it from the stack; otherwise, if the fulfilling thread stalled after the pop, the waiting thread could execute an unbounded number of steps after the pop linearized, without making progress, and other operations could linearize in-between. A push operation therefore pushes a data node regardless of the state of the stack. If the previous top-of-stack node was a reservation, the adjacent nodes then "annihilate each other": any thread that finds a data node and an underlying reservation at the top of the stack attempts to write the address of the former into the latter, and then pop both nodes from the stack.

Nonblocking dual data structures have proven quite useful in practice. In particular, the Executor framework of Java 6 uses dualstacks and dualqueues to replace the lock-based task pools of Java 5, resulting in improvements of 2–10× in the throughput of thread dispatch [Scherer et al., 2009].

8.8 NONBLOCKING ELIMINATION

In Section 5.4 we described the notion of *elimination*, which allows operations in a fan-in tree not only to combine (so that only one thread continues up the tree), but to "cancel each other out," so that neither thread needs to proceed.

Hendler et al. [2004] use elimination in a nonblocking stack to "back off" adaptively in the wake of contention. As in a Treiber stack (Section 8.1), a thread can begin by attempting a CAS on the top-of-stack pointer. When contention is low, the CAS will generally succeed. If it fails, the thread chooses a slot in (some subset of) the elimination array. If it finds a matching operation already parked in that slot, the two exchange data and complete. If the slot is empty, the thread parks its own operation in it for some maximum time t, in hopes that a matching operation will arrive. Modifications to a slot—parking or eliminating—are made with CAS to resolve races among contending threads.

If a matching operation does not arrive in time, or if a thread finds a nonmatching operation in its chosen slot (e.g., a push encounters another push), the thread attempts to access the top-of-stack pointer again. This process repeats—back and forth between the stack and the elimination array—until either a push/pop CAS succeeds in the stack or an elimination CAS succeeds in the array. If recent past experience suggests that contention is high, a thread can go directly to the elimination array at the start of a new operation, rather than beginning with a top-of-stack CAS.

To increase the odds of success, threads dynamically adjust the subrange of the elimination array in which they operate. Repeated failure to find a matching operation within the time interval t causes a thread to use a smaller prefix of the array on its next iteration. Repeated failure to eliminate successfully given a matching operation (as can happen when some *other* operation manages to eliminate first) causes a thread to use a larger prefix. The value of t, the overall size of the array, the number of failures required to trigger a subrange change, and the factor by which it changes can all be tuned to maximize performance.

Similar techniques can be used for other abstractions in which operations may "cancel out." Scherer et al. [2005] describe an *exchange channel* in which threads must "pair up" and swap information; a revised version of this code appears as the Exchanger class in the standard Java concurrency library.

With care, elimination can even be applied to abstractions like queues, in which operations cannot naively eliminate in isolation. As shown by Moir et al. [2005], one can delay an enqueue operation until its datum, had it been inserted right away, would have reached the head of the queue: at that point it can safely combine with any arriving dequeue operation. To determine when an operation is "sufficiently old," it suffices to augment the nodes of an M&S queue with monotonically increasing serial numbers. Each enqueue operation in the elimination array is augmented with the count found at the tail of the queue on the original (failed) CAS attempt. When the count at the *head* of the queue exceeds this value, the enqueue can safely be eliminated. This "FIFO elimination" has the nontrivial disadvantage of significantly increasing the latency of dequeue operations that encounter initial contention, but it can also significantly increase scalability and throughput under load.

8.9   UNIVERSAL CONSTRUCTIONS

In a perfect world, one could take a sequential implementation of some arbitrary data abstraction, run it through an automatic tool, and obtain a fast, nonblocking, concurrent implementation. Two of these properties (nonblocking and concurrent) are easy to obtain. Herlihy [1991]'s original paper on wait-free synchronization included a *universal construction* that would generate a wait-free concurrent implementation of any given sequential object.

In a subsequent paper, Herlihy [1993] described alternative constructions for both wait-free and lock-free implementations. The intuition behind the lock-free construction is particularly straightforward: access to the data structure is always made through a distinguished *root* pointer. A read-only operation need only dereference the pointer, find what it needs, and return. To modify the structure, however, a thread must create a *copy*, verify that it has done so atomically (by double-checking the root pointer), modify the copy, and then attempt to install the update by using LL / SC to swing the root pointer from the old version to the updated copy. This construction is quite efficient for small data structures. It also works well for small changes to large trees: since versions of the tree are immutable once installed, portions that do not differ can be shared between the old version and the new. If the tree is balanced, the overall cost of an update is only $\Omega(\log n)$. Herlihy exploits this observation to build an efficient concurrent *skew heap* (a priority queue implemented as an approximately balanced tree).

Over the course of the 1990s, many papers were published proposing new universal constructions. Probably the best known of these is the work of Shavit and Touitou [1995], who coined the term *software transactional memory* (STM). Inspired by the hardware proposal of Herlihy and Moss [1993], Shavit and Touitou-style STM is essentially a lock-free software implementation of a k-word compare_and_swap operation, for arbitrary k. Most subsequent STM systems have extended this functionality to accommodate *dynamic* operations, in which the full set of locations to be accessed is not known in advance. We will consider transactional memory in more detail in Chapter 9. Most STM systems use locks "under the hood"; a few, however, are lock free or obstruction free.

CHAPTER 9

Transactional Memory

Transactional memory (TM) is among the most active areas of recent synchronization research, with literally hundreds of papers published over the past ten years. The current chapter attempts to outline the shape of the TM design space, the current state of the art, and the major open questions. For further details, readers may wish to consult the encyclopedic lecture of Harris et al. [2010].

At its core, TM represents the fusion of two complementary ideas: first, that we should raise the level of abstraction for synchronization, allowing programmers to specify *what* should be atomic without needing to specify *how* to make it atomic; second, that we should employ (at least in many cases) an underlying implementation based on speculation. The user-level construct is typically simply an `atomic` label attached to a block of code. The speculative implementation allows transactions (executions of atomic blocks) to proceed in parallel unless and until they *conflict* with one another (access the same location, with at least one of them performing a write). At most one conflicting transaction is allowed to continue; the other(s) *abort*, roll back any changes they have made, and try again.

Ideally, the combination of `atomic` blocks and speculation should provide (much of) the scalability of fine-grain locking with (most of) the simplicity of coarse-grain locking, thereby sidestepping the traditional tradeoff between clarity and performance. The combination also offers a distinct semantic advantage over lock-based critical sections, namely *composability*.

An atomicity mechanism is said to be composable if it allows smaller atomic operations to be combined into larger atomic operations without the possibility of introducing deadlock. Critical sections based on fine-grain locks are not composable: if operations are composed in different orders in different threads, they may attempt to acquire the same set of locks in different orders, and deadlock can result. Speculation-based implementations of `atomic` blocks break the "irrevocability" required for deadlock (Section 3.1.1): when some transactions abort and roll back, others are able to make progress.

As noted at the end of Chapter 2, TM was originally proposed by Herlihy and Moss [1993]. A similar mechanism was proposed concurrently by Stone et al. [1993], and precursors can be found in the work of Knight [1986] and Chang and Mergen [1988]. Originally perceived as too complex for technology of the day, TM was largely ignored in the hardware community for a decade. Meanwhile, as mentioned at the end of Chapter 8, several groups in the theory community were exploring the notion of *universal constructions* [Anderson and Moir, 1999, Barnes, 1993, Herlihy, 1993, Israeli and Rappoport, 1994, Shavit and Touitou, 1995, Turek et al., 1992], which could transform a correct sequential data structure, mechanically, into a correct concurrent data structure. Shortly after the turn of the century, breakthrough work in both hardware [Martínez and Torrellas, 2002, Rajwar and Goodman, 2002, 2001] and software [Fraser and Harris, 2007, Harris and Fraser, 2003, Herlihy et al., 2003b] led to a resurgence of interest in TM. This resurgence was fueled, in part, by the move to multicore processors, which raised profound concerns about the ability of "ordinary" programmers to write code (correct code!) with significant amounts of exploitable thread-level parallelism.

Much of the inspiration for TM, both originally and more recently, has come from the database community, where transactions have been used for many years. Much of the theory of database transactions was developed in the 1970s [Eswaran et al., 1976]. Haerder and Reuter [1983] coined the acronym *ACID* to describe the essential semantics: a transaction should be

atomic – it should happen exactly once (or not at all)

consistent – it should maintain all correctness properties of the database

isolated – its internal behavior should not be visible to other transactions, nor should it see the effects of other transactions during its execution

durable – once performed, its effects should survive system crashes

These same semantics apply to TM transactions, with two exceptions. First, there is an (arguably somewhat sloppy) tendency in the TM literature to use the term "atomic" to mean both atomic and isolated. Second, TM transactions generally dispense with durability. Because they may encompass as little as two or three memory accesses, they cannot afford the overhead of crash-surviving disk I/O. At the same time, because they are intended mainly for synchronization among threads of a single program (which usually live and die together), durability is much less important than it is in the database world.

Composability

We also used the term "composability" in Section 3.1.2, where it was one of the advantages of linearizability over other ordering criteria. The meaning of the term there, however, was different from the meaning here. With linearizability, we wanted to ensure, locally (i.e., on an object-by-object basis, without any need for global knowledge or control), that the orders of operations on different objects would be mutually consistent, so we could compose them into a single order for the program as a whole. In transactional memory, we want combine small operations (transactions) into larger, still atomic, operations. In other words, we're now composing operations, not orders.

Interestingly, the techniques used to implement linearizable concurrent objects do not generally support the creation of atomic composite operations: a linearizable operation is designed to be visible to all threads before it returns to its caller; its effect can't easily be delayed until the end of some larger operation. Conversely, the techniques used to implement composable transactions generally involve some sort of global control—exactly what linearizability was intended not to need.

Given a correct sequential implementation of a data structure (a tree-based set, for example), TM allows the author of a parallel program to reuse the sequential code, with guaranteed correctness, in an almost trivial fashion:

```
class pset                              pset.insert(x : item):
    set S                                   atomic
                                                S.insert(x)
bool pset.lookup(x : item):
    atomic                              pset.remove(x : item):
        return S.lookup(x)                  atomic
                                                S.remove(x)
```

Moreover, unlike lock-based critical sections, transactions can safely be composed into larger atomic operations:

```
P, Q : pset
...
atomic
    if P.lookup(x)
        P.remove(x)
        Q.insert(x)
```

Here the fact that P.lookup, P.remove, and Q.insert contain nested transactions is entirely harmless. Moreover, if some other thread attempts a concurrent, symmetric move from Q to P, deadlock can never result.

The original intent of TM was to simplify the construction of library-level concurrent data abstractions, with relatively small operations. Current hardware (HTM) and (to a lesser extent) software (STM) implementations serve this purpose well. How much larger transactions can get before they conflict too often to scale is still an open question.

The following two sections consider software and hardware transactions in turn; the third takes a closer look at challenges—many of them initially unanticipated—that have complicated the development of TM, and may yet determine the degree of its success.

While early STM implementations were provided simply as library packages—with entry points to begin a transaction, read or write a shared location transactionally, and (attempt to) commit a transaction—experience suggests that such libraries are too cumbersome for most programmers to use [Dalessandro et al., 2007]. We assume through the rest of this chapter that TM is embedded in a programming language, and that all necessary hooks (including instrumentation of STM loads and stores) are generated by the compiler.

9.1 SOFTWARE TM

If two TM implementations provide the same functionality—one in hardware and the other in software—the hardware version will almost certainly be faster. Software implementations have other advantages, however: they can run on legacy hardware, they are more flexible (extensible), and they can provide functionality that is considered too complex to implement in hardware.

As of this writing, hardware TM systems are just beginning to reach the consumer market. The majority of research over the past decade has taken place in a software context.

9.1.1 DIMENSIONS OF THE STM DESIGN SPACE

The variety of STM designs has proven to be something of a surprise: researchers a decade ago did not anticipate how large the design space would be. Principal design dimensions include the following:

Progress guarantees – Most of the early universal constructions were nonblocking, and many of the original STM systems were likewise. The OSTM (object-based STM) of Fraser's thesis work was lock free [Fraser, 2003, Fraser and Harris, 2007]; several other systems have been obstruction free [Harris and Fraser, 2003, Herlihy et al., 2003b, Marathe and Moir, 2008, Marathe et al., 2005, 2006, Tabba et al., 2009]. Over time, however, most groups have moved to blocking implementations in order to obtain better expected-case performance.

Buffering of speculative updates – To be able to roll back aborted transactions, a TM system must maintain both the old and new versions of speculatively modified data. The two most common options are sometimes referred to as "undo logging" [Harris et al., 2006, Saha et al., 2006a], in which changes are made "in place," but old values are buffered for write-back on abort, and "redo logging" [Dice et al., 2006, Harris and Fraser, 2003], in which new values are buffered for write-back on commit. In either case, a secondary decision must be made regarding the granularity (byte, word, cache line, …) at which to buffer changes. In a system with redo logs, load instructions must be modified to first check the redo log, to make sure the transaction sees its own writes. Some object-based systems use a third, "cloning" option, in which multiple copies of a language-level object are globally visible, and threads choose among them based on metadata that tracks commits and aborts [Fraser and Harris, 2007, Herlihy et al., 2003b, Marathe et al., 2005, 2006, Tabba et al., 2009].

Access tracking and conflict resolution – When two transactions conflict, a TM system must ensure that they do not both commit. Some systems are *eager*: they notice as soon as a location already accessed in one transaction is accessed in a conflicting way by a second transaction. Other systems are *lazy*: they delay the resolution of conflicts until one of the transactions has finished execution and is ready to commit. A few systems are *mixed*: they resolve write-write conflicts early but read-write conflicts late [Dragojević et al., 2009, Scott, 2006, Shriraman and Dwarkadas, 2009].

To detect conflicts, a TM system must track the accesses performed by each transaction. In principle, with lazy conflict resolution, one could log accesses locally in each thread, and intersect, at commit time, the logs of transactions that overlap in time. RingSTM [Spear et al., 2008a], indeed, does precisely this. More commonly, TM systems employ some sort of shared *metadata* for access tracking. Some object-oriented systems include metadata in the header of each object. Most STM systems, however, use a hash function keyed on the address of the accessed location to index into a a global table of "ownership" records (Orecs). By ignoring the low bits of the address when hashing, we can arrange for the bytes of a given block (word, cache line, etc.) to share the same Orec. Given many-to-one hashing, a single Orec will also, of course, be shared by many blocks: this *false sharing* means that logically independent transactions will sometimes appear to conflict, forcing us to choose between them.

Lazy and mixed conflict resolution have the advantage that readers can avoid updating metadata to make themselves visible to writers. A system that skips these updates is said to have *invisible readers*. Because metadata updates tend to induce cache misses, eliminating them can dramatically improve the performance of read-only or read-mostly transactions.

Validation – It is straightforward to demonstrate that an STM system will guarantee strict serializability if it never commits conflicting transactions that overlap in time. In a system with invisible readers, we commonly distinguish between *validation* and the rest of conflict resolution. In a given transaction A, validation serves to ensure that no other transaction B has made in-place updates (or committed updates) to locations read by A. When a read-only transaction (one that modifies no shared locations) completes its execution, successful validation is all that it requires in order to commit. When a writer transaction completes, it must also make its updates visible to other threads. In an Orec-based system with a redo-log (e.g., TL2 [Dice et al., 2006]), a transaction will typically lock the Orecs of all locations it wishes to modify, validate, write back the contents of its redo log, and then unlock the Orecs.

In a system with lazy conflict resolution, validation must also be performed on occasion *during* transaction execution—not just at the end. Otherwise a transaction that has read mutually inconsistent values of memory locations (values that could not logically have been valid at the same time) may perform operations that would never occur in any sequential execution, possibly resulting in faults (e.g., divide-by-zero), infinite loops, nontransactional (uninstrumented) stores to shared addresses, or branches to nontransactional (uninstrumented) code. A maximally pessimistic system may choose to validate immediately after every shared read; such a system is said to preserve *opacity* [Guerraoui and Kapałka, 2008]. A more optimistic system may delay validation until the program is about to execute a "dangerous" operation; such a system is said to be *sandboxed* [Dalessandro and Scott, 2012].

Contention management – To resolve conflicts among transactions, we must choose which of the contenders should continue and which should abort or wait. (Waiting may be a viable option if the transaction that continues is a reader, working with the original value of the conflicting datum, and we allow it to finish before the writer continues. Waiting may also be viable if there is a possibility that the transaction that continues may ultimately fail to commit.) If certain atomic blocks frequently access the same data, we may wish to alter system scheduling to reduce the likelihood that those blocks will execute concurrently. We may also wish to make choices, over time, that ensure some measure of fairness (Section 3.2.2) among transactions or threads. These concerns are all issues of liveness (Section 3.2). In some systems—particularly those with eager conflict resolution—it may be useful, from a software engineering perspective, to separate liveness from safety (correctness) concerns, and address it in a separate *contention management* module.

With the exception of progress guarantees, we will discuss each of these design space dimensions in its own subsection below. Readers who are interested in exploring the alternatives may wish to download the RSTM suite [RSTM], which provides a wide variety of interchangeable STM "back ends" for C++.

The design space dimensions are largely but not fully orthogonal. When transactions conflict, there is no way for a writer to defer to a reader it cannot see: invisible readers reduce the flexibility of contention management. In a similar vein, private undo logs (not visible to other threads) cannot be used in a nonblocking system, and private access logs cannot be used for eager conflict resolution. Perhaps most important, there is no obvious way to combine in-place update (undo logs) with lazy conflict resolution: Suppose transaction A reads x, transaction B writes x (speculatively, in place), and transaction A is the first to complete. Without knowing whether A's read occurred before or after B's write, we have no way of knowing whether it is safe to commit A.

9.1.2 BUFFERING OF SPECULATIVE STATE

Transactions mediate access to shared data, but they may also access thread-private data. On an abort, these private data may need to be rolled back. In principle, one could treat *all* data as shared, but this would induce unnecessary overhead in the private case. Variables that are declared within the lexical scope of an atomic block require no load/store instrumentation: because their lifetimes are bounded by that of the transaction (and their storage is typically above the value of the stack pointer that will be restored on abort), there is nothing to roll back to. Variables that are private but were created outside the transaction will need to be rolled back. Because no other thread can modify them, it suffices to checkpoint their values at the beginning of the transaction, and simply restore them on abort: no per-access instrumentation is required.

Compilers may use conservative static analysis to distinguish between private and shared data. Any access that might be to shared data must be instrumented. Any pre-existing private variable that might be modified in the transaction must be checkpointed (and restored on abort)—unless *every* store that might access it is instrumented.

Dynamically allocated data require special treatment. Malloc and free (or their equivalents) can be made transaction-safe. Data malloced in an aborted transaction must be freed. Data freed in a transaction must be kept in limbo until we know whether the transaction will commit. With lazy conflict resolution, freed data is probably best kept in limbo until all potentially concurrent transactions have completed.

As noted above, most TM systems are prone to *false sharing*—accesses to disjoint data that are covered by the same metadata. Here the mismatch between access and metadata granularity may lead to unnecessary aborts. Granularity is also a problem for redo and undo logging. Suppose that transaction A stores to byte b of word w, and that nontransactional code stores to byte c. If the TM system logs accesses at full-word granularity, then the write-back of w on commit (in a redo-log system) or abort (in an undo-log system) may overwrite the nontransactionally updated value of c. Note that the example entails no data race between transactional and nontransactional code: the problem is one of implementation. To maintain correctness, either the compiler must ensure (via data structure layout) that transactional and nontransactional instructions never access different pieces of a logging-granularity block, or the TM system must log at the finest possible granularity. One possible optimization is to log at word granularity, but tag each logged word with per-byte dirty bits.

9.1.3   ACCESS TRACKING AND CONFLICT RESOLUTION

Two concurrent transactions conflict if they both access the same location, and at least one of them writes it. Conflicts can be resolved by aborting either transaction, or (in some cases) by stalling the transaction that makes its access second. Without prior knowledge of read and write sets (and without enforcing global mutual exclusion), a TM system must speculate (and be prepared to roll back) at least up to the first conflict. It can speculate beyond that point, but then write-write conflicts among transactions will eventually force all but one to abort, and read-write conflicts will force the reader(s) to abort if a writer commits first.

Generalizing the Notion of Conflict

Write-write and read-write conflicts interfere with serializability because the accesses don't commute: they can lead to different results if performed in the opposite order. Concurrent reads of the same location by different transactions do not interfere with serializability because they *do* commute: they produce the same results if performed in the opposite order.

With this observation in mind, one can raise the level of abstraction, generalize the notion of conflict, decrease the overhead of conflict detection, and reduce the frequency with which transactions conflict. We have already seen an example with malloc and free. Unless these are treated as a special case, loads and stores to locations in the memory manager (e.g., a global free list) are likely to be seen as conflicts, even though operations on separate blocks of memory are logically independent. By special-casing malloc and free, we essentially add them to the list of primitive operations (along with load and store), and we note that (as long as the memory manager is properly synchronized internally), operations on separate blocks commute at the level of program semantics, even if they result in different memory contents (different blocks in different places) at the level of the implementation.

Herlihy and Koskinen [2008] have proposed a *transactional boosting* mechanism to add abstractions to the TM system, so that conflicts on a set abstraction, for example, might be tracked at the level of add, remove, and lookup operations, rather than individual loads and stores. Two add operations would then commute—and fail to cause a conflict—even if they resulted in a different concrete representation of the (abstract) set in memory. Boosting requires that every operation have an *inverse* that can be used to undo its effects. It also benefits from an asymmetric generalization of the notion of commutativity [Koskinen et al., 2010].

There is a fundamental tradeoff between eager and lazy conflict resolution. Let us say that a conflict *occurs* at the time of the second of a pair of conflicting accesses. The conflict is *resolved* when the TM system chooses which transaction will be given the chance to commit. (In some systems, there may be a point in-between at which the conflict is *detected*, but not yet resolved.) With lazy conflict resolution, if transaction A eventually aborts, any work it performs after the original occurrence of a conflict might be seen as "egregiously wasted": in hindsight, given the conflict, there is no way it could have been productive. With eager conflict resolution, however, we may not know, at the time of the conflict, what other conflicts may arise in the future. If transaction A aborts in deference to transaction B, but B subsequently aborts in deference to some other transaction C, then in hindsight there was no point in aborting A: its work has been wasted just as egregiously as it was in the lazy case. There is no way, in general, to tell whether eager or lazy conflict resolution will result in more wasted work. Mixed resolution exploits the observation that in a redo-log-based system, both transactions can commit in the wake of a read-write conflict, if the reader does so first.

As noted above, invisible readers avoid the problem of "turning readers into writers" by forcing them to update metadata. By introducing asymmetry between readers and writers, however, we force conflict resolution to be performed by readers. Moreover, read-write conflicts in which the read occurs first must be detected after the fact: a reader must revisit and *validate* its reads before committing (Section 9.1.4). An intermediate option, explored in the SkySTM system [Lev et al., 2009a], is to use a *scalable non-zero indicator* (SNZI) [Ellen et al., 2007] to indicate the presence of one or more readers, without recording their individual identities. SNZI uses a tree-based representation to reduce contention for the reader count, thereby reducing the overhead of read-only transactions while still allowing writers to detect the existence of conflicts with one or more earlier readers.

9.1.4 VALIDATION

As described in Section 3.1.2, two-phase locking provides a straightforward way to ensure serializability. Each transaction, as it runs, acquires a reader-writer lock (in read or write mode as appropriate) on the Orec of every location it wishes to access. (This implies eager conflict detection.) If an Orec is already held in an incompatible mode, the transaction stalls, aborts, or (perhaps) kills the transaction(s) that already hold the lock. (This implies eager conflict *resolution*. To avoid deadlock, a transaction that stalls must do so provisionally; if it waits too long it must time out and abort.) If all locks are held from the point of their acquisition to the end of the transaction, serializability is ensured. As described in the previous subsection, SNZI can be used to reduce the contention and cache misses associated with lock updates by readers, at the expense, in a writer, of not being able to identify *which* transactions have already acquired an Orec in read mode.

To implement invisible readers, we can use sequence locks to replace the reader-writer locks on Orecs. A reader makes no change to the (shared) lock, but does keep a (private) record of the *value* of the lock at the time it reads a covered location. The record of lock values constitutes a *read log*, analogous to the *write log* already required for redo on commit or undo on abort. Using its read log, a transaction can *validate* its reads by double-checking the values of Orec locks: if a lock has changed, then some other transaction has acquired the Orec as a writer, and the covered data can no longer be assumed to be consistent; the reader must abort.

To reduce the impact of false sharing, a transaction can choose to keep the values of loaded locations in its read log, instead of—or in addition to—the values of Orec locks. It can then perform *value-based validation* [Ding et al., 2007, Olszewski et al., 2007], verifying that previously-read locations still (or again) contain the same values. Some mechanism—typically a check of Orec lock values—must still be used, of course, to guarantee that the verified values are all present at the same time.

In the degenerate case, Dalessandro et al. [2010c] use a single global Orec to provide this guarantee. Their "NOrec" system allows a read-only transaction to validate—and commit—without acquiring any locks: the transaction reads the global sequence lock, uses value-based-validation to verify the consistency of all read locations, and then double-checks the sequence lock to make sure that no other transaction committed writes during the validation. As in any system with invisible readers, they employ a redo log rather than an undo log, and they validate during the transaction immediately after every shared read or (with sandboxing [Dalessandro and Scott, 2012]) immediately before every "dangerous" operation. NOrec forces transactions to write back their redo logs one at a time, in mutual exclusion, but it allows them to *create* those logs—to figure out what they want to write—in parallel. As of early 2013, no known STM system consistently outperforms NOrec for realistic workloads on single-chip multicore machines, though both TML [Dalessandro et al., 2010a] and FastLane [Wamhoff et al., 2013] are better in important cases.

Avoiding Redundant Validation

If a transaction reads n different locations, and validates its previous reads after each new read (to guarantee opacity), Orecs will need to be consulted $O(n^2)$ times over the course of the transaction. For large transactions, the overhead of this *incremental validation* can be prohibitive. To reduce it, Spear et al. [2006] observe that in a redo-log-based system, transaction T must validate its reads only if some other transaction has committed writes since T's last validation. Absent a very large number of cores and a very high transaction completion rate, many validations can be avoided if we maintain a global count of the number of committed writers, and elide validation whenever the count has not changed. NOrec uses the global sequence lock to effect the same optimization.

Time-based Validation

By validating its previous reads immediately after reading a new location x, transaction T ensures that even if x has been modified very recently, all of T's work so far is still valid, because the other locations have not been modified since they were originally read. An alternative approach, pioneered by the TL2 system of Dice et al. [2006], is to verify that the newly read location x has not been modified since T began execution. That is, instead of ensuring that all of T's work so far is correct as of the current moment, we ensure that it is correct *as of T's start time*. To implement this approach, TL2 employs a global "clock" (actually, just a global count of committed transactions). It then augments each Orec with a *version number* that specifies the value of the global clock as of the most recent write to any location covered by the Orec. At the beginning of each transaction, TL2 reads and remembers the global clock. On each read, it verifies that the version number in the corresponding Orec is less than or equal to the remembered clock value. If not, the transaction aborts.

If a read-only transaction completes its execution successfully, we know its behavior is correct as of its start time. No additional work is necessary; it trivially commits. A writer transaction, however, must validate its read set. It locks the Orecs of all locations it wishes to write, atomically increments the global clock, checks the version numbers of (the Orecs of) all locations it has read, and verifies that all are still less than its start time (so the covered locations have not been modified since). If it is unable to acquire any of the Orecs for the write set, or if any of the Orecs for the read set have too-recent version numbers, the transaction aborts. Otherwise, it writes back the values in its redo log and writes the (newly incremented) global clock value into each locked Orec. By colocating the lock and version number in a single word, TL2 arranges for these writes to also unlock the Orecs.

When a transaction T in TL2 reads a location x that has been modified since T's start time, the transaction simply aborts. Riegel et al. [2006] observe, however, that just as a writer transaction must validate its reads at commit time, effectively "extending" them to its completion time, a reader or writer transaction can update its reads incrementally. If T began at time t_1, but x has been modified at time $t_2 > t_1$, T can check to see whether any previously read location has been modified since t_2. If not, T can pretend it began at time t_2 instead of t_1, and continue. This *extensible timestamp* strategy is employed in the TinySTM system of Felber et al. [2008], which has invisible readers but eager conflict detection. It is also used in SwissTM [Dragojević et al., 2009], with mixed conflict detection, and NOrec [Dalessandro et al., 2010c], with lazy detection.

Bloom Filters as an Alternative to Read Logs

While NOrec is easily described as a single-Orec STM system with value-based validation, its inspiration came not from Orec-based systems, but from an earlier system known as RingSTM [Spear et al., 2008a]. RingSTM uses redo logs and lazy conflict detection, but its validation is based not on the write log and a full read set, but rather on *Bloom filter* approximations of these.

The behavior of RingSTM is most easily understood by imagining an unbounded global list whose entries represent committed transactions, in serialization order. Each entry of the list contains the write set of the transaction, summarized as a Bloom filter. When it begins execution, a transaction reads and remembers a pointer to the head of the list. While running, it builds Bloom filters that represent its reads and writes. To validate, it intersects its read and write filters with the write filters of any transactions that have been added to the global commit list since the last validation (or the start of the transaction). If all intersections are empty, the transaction updates its pointer to the head of the list. If some intersection is nonempty, a conflict (true or false) has occurred, and the transaction aborts. To commit, a transaction reads the current head pointer, validates, and then uses CAS to add its own Bloom filter to the head of the list. A successful CAS constitutes serialization. The transaction then writes back its redo log, waits until all previously-committed transactions have finished doing the same (as indicated in their entries of the global list) and finally marks its own write-back as finished.

While an unbounded global list is clearly impractical, we really only need the portion that was created after the start time of the oldest running transaction. In practice, the list is replaced with a bounded circular buffer (a ring—hence the name), and modest extra checks avoid any problems with wrap-around.

In comparison to NOrec, RingSTM has higher costs for load/store instrumentation, but lower costs for validation (at least in large transactions). It also allows concurrent write-back.

9.1.5   CONTENTION MANAGEMENT

In a system with lazy conflict resolution, contention management is more or less straightforward: since one transaction is ready to commit, and cannot encounter further problems at this point, it should generally win out over any partially-completed transactions with which it conflicts [Spear et al., 2009a]. In particular, letting it do so ensures livelock freedom: whenever a conflict is re-solved, some transaction is guaranteed to commit, and the system as a whole makes progress. Starvation may still be a problem, especially for long-running transactions that read a large num-ber of transactions. Spear et al. report good results with a simple heuristic: any transaction that aborts repeatedly will eventually acquire a global lock that blocks other transactions from com-mitting, and makes completion of the previously starving transaction inevitable.

It is much less clear what strategy to use for systems with eager conflict resolution. Possi-bilities include favoring the transaction that started earlier, or has read or written a larger number

---

Bloom Filters

For readers not familiar with the notion, a Bloom filter [Bloom, 1970] is a bit vector representation of a set that relies on one or more hash functions. Bit i of the vector is set if and only if for some set member e and some hash function h_j, $h_j(e) = i$. Element e is inserted into the vector by setting bit $h_j(e)$ for all j. The lookup method tests to see if e is present by checking all these bits. If all of them are set, lookup will return true; if any bit is unset, lookup will return false. These conventions allow false positives (an element may appear to be present when it is not), but not false negatives (a present element will never appear to be absent). In the basic implementation, deletions are not supported.

Note that Bloom filters do not introduce a qualitatively different problem for TM: Orec-based STM systems already suffer from false sharing. The actual rate of false positives in RingSTM depends on the application and the choice of Bloom filter size.

of shared locations, or has been forced to abort more often, or has already killed off a larger number of competitors. In general, these strategies attempt to recognize and preserve the investment that has already been made in a transaction. There was a flurry of papers on the subject back around 2005 [Guerraoui et al., 2005a,b, Scherer III and Scott, 2005a,b], and work continues to be published, but no one strategy appears to work best in all situations.

9.2   HARDWARE TM

While (as we have seen) TM can be implemented entirely in software, hardware implementations have several compelling advantages. They are faster, of course, at least for equivalent functionality. Most can safely (and speculatively) call code in unmodified (uninstrumented) binary libraries. Most guarantee that transactions will serialize not only with other transactions, but also with individual (non-transactional) loads, stores, and other atomic instructions. (This property is sometimes known as *strong atomicity* or *strong isolation* [Blundell et al., 2005].) Finally, most provide automatic, immediate detection of inconsistency, eliminating the need for explicit validation.

Most of the design decisions discussed in Section 9.1, in the context of STM, are relevant to HTM as well, though hardware may impose additional restrictions. Contention management, for example, will typically be quite simple, or else deferred to software handlers. More significantly, buffer space for speculative updates is unlikely to exceed the size of on-chip cache: transactions that exceed the limit may abort even in the absence of conflicts. Transactions may also abort for any of several "spurious" reasons, including context switches and external interrupts.

In any new hardware technology, there is a natural incentive for vendors to leverage existing components as much as possible, and to limit the scope of changes. Several HTM implementations have been designed for plug-compatibility with traditional cross-chip cache coherence protocols. In the IBM Blue Gene/Q [Wang et al., 2012], designers chose to use an unmodified processor core, and to implement HTM entirely within the memory system.

To accommodate hardware limitations, most HTM systems—and certainly any commercial implementations likely to emerge over the next few years—will require software backup. In the simple case, one can always fall back to a global lock. More ambitiously, we can consider *hybrid* TM systems in which compatible STM and HTM implementations coexist.

In the first subsection below we discuss aspects of the TM design space of particular significance for HTM. In Section 9.2.2 we consider *speculative lock elision*, an alternative ABI that uses TM-style speculation to execute traditional lock-based critical sections. In 9.2.3 we consider alternative ways in which to mix hardware and software support for TM.

9.2.1   DIMENSIONS OF THE HTM DESIGN SPACE

HTM work to date includes a large number of academic designs and half a dozen commercial implementations. We mention the former in the sections below, but focus mostly on the latter. They include the Azul Systems Vega 2 and 3 [Click, 2009]; the experimental Sun/Oracle Rock processor [Dice et al., 2009]; three independently developed systems—for the Blue Gene/Q [Wang et al., 2012], zEC12 mainframe [Jacobi et al., 2012], and Power 8 [IBM, 2012]—from IBM, and the Transactional Synchronization Extensions (TSX) of Intel's "Haswell" processor [Intel, 2012].

ABI

Most HTM implementations include instructions to start a transaction, explicitly abort the current transaction, and (attempt to) commit the current transaction. (In this chapter, we refer to these, generically, as tx_start, tx_abort, and tx_commit.) Some implementations include additional instructions, e.g., to suspend and resume transactions, or to inspect their status.

While a transaction is active, load and store instructions are considered speculative: the hardware automatically buffers updates and performs access tracking and conflict detection. Some systems provide special instructions to access memory *nonspeculatively* inside of a transaction—e.g., to spin on a condition or to save information of use to a debugger or performance analyzer.[1] Because these instructions violate isolation and/or atomicity, they must be used with great care.

On an abort, a transaction may retry automatically (generally no more than some fixed number of times), retry the transaction under protection of an implicit global lock, or jump to a software handler that figures out what to do (e.g., retry under protection of a software lock). In Intel's RTM (Restricted Transactional Memory—part of TSX), the address of the handler is an argument to the tx_start instruction. In IBM's z and Power TM, tx_start sets a condition code, in the style of the Posix setjmp routine, to indicate whether the transaction is beginning or has just aborted; this code must be checked by the following instruction. With either style of abort delivery, any speculative updates performed so far will be discarded.

In the IBM Blue Gene/Q, HTM operations are triggered not with special instructions, but with stores to special locations in I/O space. Conflicts raise an interrupt, which is fielded by the OS kernel.

Buffering of Speculative Updates

There are many ways in which a processor can buffer speculative updates. Herlihy and Moss's original proposal called for a special *transactional cache* located alongside the L1 data cache. Sun's prototype Rock processor (never brought to market) allowed only a very small number of speculative stores, which it kept in the core's store buffer [Dice et al., 2009]. A few academic systems have proposed keeping logs in virtual memory [Ananian et al., 2005, Bobba et al., 2008, Yen et al., 2007], but by far the most common approach is to buffer speculative updates at some level of the normal cache hierarchy, and to "hide" them from the coherence protocol until the transaction is ready to commit. In some systems (e.g., Blue Gene/Q and the original Herlihy and Moss proposal), the cache holds both the original and the speculative version of a line, but most systems

---

[1]Among commercial machines (as of this writing), z TM provides nontransactional stores (ordered at commit or abort time), but not loads. Sun's Rock provided both, with stores again ordered at commit/abort. Intel's TSX provides neither. Power TM allows transactions to enter a "suspended" state (see page 160) in which loads and stores will happen immediately and "for real." Blue Gene/Q facilities can be used to similar ends, but only with kernel assistance. On both the Power 8 and Blue Gene/Q, the programmer must be aware of the potential for paradoxical memory ordering.

buffer only the speculative version; the original can always be found in some deeper level of cache or memory.

Whatever the physical location used to buffer speculative updates, there will be a limit on the space available. In most HTM systems, a transaction will abort if it overflows this space, or exceeds the supported degree of associativity (footnote, page 13). Several academic groups have proposed mechanisms to "spill" excess updates to virtual memory and continue to execute a hardware transaction of effectively unbounded size [Blundell et al., 2007, Ceze et al., 2006, Chuang et al., 2006, Chung et al., 2006, Rajwar et al., 2005, Shriraman et al., 2010], but such mechanisms seem unlikely to make their way into commercial systems anytime soon.

In addition to the state of memory, a TM system must consider the state of in-core resources—registers in particular. In most HTM systems, tx_begin checkpoints all or most of the registers, and restores them on abort. In a few systems (including Blue Gene/Q and the Azul Vega processors [Click, 2009]), software must checkpoint the registers prior to tx_begin, and restore them manually on abort.

Access Tracking and Conflict Resolution

By far the most common way to identify conflicts in an HTM system is to leverage the existing cache coherence protocol. Consider a basic MESI protocol of the sort described in Section 2.1.2. Suppose, in addition to the tag bits used to indicate whether a line is modified, exclusive, shared, or invalid, we add an additional *speculative* bit. The speculative modified state indicates a line that

---

Buffering in the Cache

Though most commercial HTM implementations buffer speculative updates (and tag speculative reads) in the cache hierarchy, they do so in different ways. Azul's implementation was entirely in the (private) L1 caches; its shared L2 was oblivious to speculation. Blue Gene/Q, by contrast, keeps its buffering and tags in a 32 MB L2 cache, with very little support from the L1 caches (each of which is shared by as many as 4 threads). In so-called "short-running" mode, a transaction must bypass the L1 on every access (suffering an L1 miss penalty), so that speculative accesses can be tracked in the L2, and loads can see the stores. Alternatively, in "long-running" mode, the operating system flushes the L1 at the beginning of a transaction, and then manipulates virtual-to-physical mappings so that separate threads will use separate physical addresses in the L1 (the bits in which these addresses differ are stripped by hardware on the L1-to-L2 path).

The zEC12 takes yet another approach. It tracks and buffers speculative accesses in the L1, but both the L1 and L2 are normally write-through. To hide speculative stores from the L2 and L3, designers introduced a 64-cache line (8 KB) *gathering store cache* on the L1-to-L2 path. During a transaction, stores are held in the gathering cache until commit time; stores that target the same cache line are coalesced. Given that the L3 is shared by 6 cores, each of which can (in bursts) execute two stores (and 5 other instructions) per cycle at 5.5 GHz, the gathering cache serves the important additional function of reducing incoming bandwidth to the L3, even during normal operation.

Like the Azul machines, Intel's Haswell processor performs access tracking and update buffering in the L1 cache, at cache line granularity. Conflicts are detected using the existing coherence protocol. As of this writing, IBM has yet to reveal implementation details for the Power 8.

has been written in the current transaction; the speculative exclusive and shared states indicate lines that have been read (but not written) in the current transaction. Any incoming coherence message requesting shared access to a line in speculative modified state constitutes a conflict, as does any request for exclusive access to a speculative line of any kind.

Most HTM systems perform eager conflict detection. The TCC proposal of Hammond et al. [2004] delays detection until a transaction is ready to commit. Shriraman et al. [2010] propose to detect conflicts as they occur, but resolve them only at commit time. Blue Gene/Q's TM hardware could in principle be used to construct a lazy system, but its software runtime was designed to resolve conflicts eagerly. Even so, it is distinctive in its ability to choose, dynamically, which transaction should continue and which should wait or abort in the event of a conflict.

In an eager system, the most straightforward response to a conflict (detected via incoming coherence traffic) is to abort the current transaction. This strategy, known as "responder loses," has the advantage of full compatibility with existing coherence protocols. After aborting, the local (responding) core can provide the conflicting line to the remote (requesting) core in precisely the same way it would on a machine without HTM.

The disadvantage of "responder loses" is the potential for livelock: if the local transaction restarts and attempts to access the conflicted line before the remote transaction has completed, the roles of requester and responder will be reversed, and the remote transaction will abort. Various contention management strategies (e.g., randomized exponential backoff) can be used to minimize the problem. In the IBM zEC12 [Jacobi et al., 2012], the coherence protocol already includes a NAK message, used to (temporarily) refuse to downgrade a line for which writeback is currently in progress. The TM system leverages this message to "stiff-arm" requests for a transactional line, in hopes of completing the current transaction before the requester tries again.

Several proposals for "unbounded" HTM use hardware-implemented Bloom filter *signatures* to summarize read and write sets that overflow the cache [Ceze et al., 2006, Shriraman et al., 2010, Yen et al., 2007]. An incoming request from the coherence protocol will trigger conflict management not only due to conflict with a speculative line in the local cache, but also due to conflict with an overflowed line, as captured by the read or write set signature. Other systems control access to overflowed data using page-level memory protections [Chuang et al., 2006] or distributed *ownership tokens* [Bobba et al., 2008].

Potential Causes of Aborts

In all TM systems, transactions must sometimes abort in response to conflicts with other transactions. Most HTM systems will also abort a transaction for various other reasons, including conflicts due to false sharing, overflow of the capacity or associativity of the speculative cache, exceptions (interrupts and faults) of various kinds, and attempts to execute instructions not supported in speculative mode. Exactly which instructions and exceptions will trigger an abort differs from machine to machine. Transactions on Sun's Rock processor were particularly fragile: most subroutine calls or mispredicted branches would end the current transaction [Dice et al., 2009].

Blue Gene/Q is particularly robust: uncached memory accesses and DMA requests are the only program behaviors that force the TM system to interrupt a transaction (the operating system will also abort the current transaction on various interrupts and faults).

Most of the proposals for "unbounded" hardware transactions address not only the possibility of inadequate space for buffered updates, but also inadequate *time*: the system generally provides a way to save the state of a transaction to memory at the end of a scheduling quantum and resume it the next time its thread gets to run. Among the HTM systems that have actually been built, only Blue Gene/Q allows transactions to survive a context switch. Most others abort the current transaction on any of the interrupts (e.g., time expiration, device completion) that might end the current quantum.

Power 8 [IBM, 2012] has a distinctive mechanism to suspend and resume the current transaction. While a transaction is suspended, loads and stores proceed nontransactionally. Most exceptions cause the current transaction to be suspended rather than aborted. The operating system can choose whether to resume the transaction after servicing the exception. If it decides to effect a context switch, a special instruction allows it to *reclaim* the processor state that was checkpointed at the beginning of the transaction. At the start of the next quantum, another special instruction allows the OS to *re-checkpoint* and resume the transaction, whereupon it will immediately abort and fall into its software handler (which can then retry, fall back to a software lock, etc., as appropriate).

For mission-critical applications, IBM's z TM [Jacobi et al., 2012] supports a special *constrained* transaction mode that is guaranteed to succeed, eventually, in hardware, with no software handler or fallback path. Constrained transactions are limited to a small total number of instructions (currently 32) with no backward branches, no subroutine calls, and a small (currently 32-byte) memory access footprint. They are intended for small data structure updates—the sorts of things for which Herlihy and Moss TM (or Oklahoma Update) was originally intended. The Advanced Synchronization Facility [AMD, 2009, Diestelhorst et al. [2010]], proposed by researchers at AMD, envisions a similar guarantee for small transactions.

9.2.2 SPECULATIVE LOCK ELISION

While proponents typically argue for TM as a way to simplify the construction of scalable parallel programs, this is not necessarily a compelling argument from the perspective of hardware vendors. Given the investment needed to add TM to an existing ISA—and to propagate it through all future generations—potential improvements in the performance of *existing* programs provide a much more compelling argument. Speculative lock elision (SLE) [Rajwar and Goodman, 2002, 2001] is a use case that provides precisely this sort of improvement. It retains the traditional lock-based programming model, but attempts to execute critical sections as transactions whenever possible. Doing so has at least two potential benefits:

1. Particularly for code with medium- to coarse-grain locks, it is common for critical sections protected by the same lock to encounter no actual conflicts. SLE may allow such critical sections to execute in parallel.

2. Even when data conflicts are relatively rare, it is common for a thread to find that a lock was last accessed on a different core. By eliding acquisition of the lock (i.e., simply verifying that it is not held), SLE may avoid the need to acquire the lock's cache line in exclusive mode. By leaving locks shared among cores, a program with many small critical sections may suffer significantly fewer cache misses.

Both of these benefits may improve performance on otherwise comparable machines. They also have the potential to increase scalability, allowing programs in which locks were becoming a bottleneck to run well on larger numbers of cores.

Azul has indicated that lock elision was the sole motivation for their HTM design [Click, 2009], and the designers of most other commercial systems, including z [Jacobi et al., 2012], Power [IBM, 2012], and TSX [Intel, 2012], cite it as a principal use case. On z, SLE is simply a programming idiom, along the following lines:

```
          really_locked := false
          tx_begin
          if failure goto handler
          read lock value              // add to transaction read set
          if not held goto cs
          abort
handler:  really_locked := true
          acquire lock
cs:       ...                          // critical section
          if really_locked goto release
          tx_commit
          goto over
release:  release lock
over:
```

This idiom may be enhanced in various ways—e.g., to retry a few times in hardware if the abort appears to be transient—but the basic pattern is as shown. One shortcoming is that if the critical section (or a function it calls) inspects the value of the lock (e.g., if the lock is reentrant, and is needed by a nested operation), the lock will appear not to be held. The obvious remedy—to write a "held" value to the lock—would abort any similar transaction that is running concurrently. An "SLE-friendly" solution would require each transaction to remember, in thread-local storage, the locks it has elided.

Power TM provides a small ISA enhancement in support of SLE: the tx_commit instruction can safely be called when not in transactional mode, in which case it sets a condition code. The idiom above then becomes:

```
                tx_begin
                if failure goto handler
                read lock value              // add to transaction read set
                if not held goto cs
                abort
     handler:   acquire lock
     cs:        ...                          // critical section
                tx_commit
                if commit_succeeded goto over
                release lock
     over:
```

Originally proposed in the thesis work of Ravi Rajwar [2002], SLE plays a significantly more prominent role in Intel's Transactional Synchronization Extensions (TSX), of which Rajwar was a principal architect. TSX actually provides two separate ABIs, called Hardware Lock Elision (HLE) and Restricted Transactional Memory (RTM). RTM's behavior, to first approximation, is similar to that of z or Power TM. There are instructions to begin, commit, or abort a transaction, and to test whether one is currently active.

On legacy machines, RTM instructions will cause an unsupported instruction exception. To facilitate the construction of backward-compatible code, HLE provides an alternative interface in which traditional lock acquire and release instructions (typically CAS and store) can be tagged with an XACQUIRE or XRELEASE prefix byte. The prefixes were carefully chosen from among codes that function as nops on legacy machines; when run on such a machine, HLE-enabled code will acquire and release its locks "for real." On a TSX machine, the hardware will refrain from acquiring exclusive ownership of the cache line accessed by an XACQUIRE-tagged instruction. Rather, it will enter speculative mode, add the lock to its speculative update set, and remember the lock's original value and location. If the subsequent XRELEASE-tagged instruction restores the original value to the same location (and no conflicts have occurred in the interim), the hardware will commit the speculation. Crucially, any loads of the lock within the critical section will see its value as "locked," even though its line is never acquired in exclusive mode. The only way for code in a critical section to tell whether it is speculative or not is to execute a (non-backward-compatible) explicit XTEST instruction.

Because an XRELEASE-tagged instruction must restore the original value of a lock, several of the lock algorithms in Chapter 4 must be modified to make them HLE-compatible. The ticket lock (Figure 4.7, page 55), for example, can be rewritten as shown in Figure 9.1. Speculation will succeed only if ns = next_ticket on the first iteration of the loop in acquire, and no other thread increments next_ticket during the critical section. Note in particular that if now_serving ≠ next_ticket when a thread first calls acquire, the loop will continue to execute until the current lock holder updates either now_serving or next_ticket, at which point HLE will abort and retry the FAI "for real." More significantly, if no two critical sections conflict, and if no aborts occur due to overflow or other "spurious" reasons, then an arbitrary number of threads can execute crit-

```
class lock
    int next_ticket := 0
    int now_serving := 0
    const int base = ...                      // tuning parameter
lock.acquire():
    int my_ticket := XACQUIRE FAI(&next_ticket)
    loop                                                        lock.release():
        int ns := now_serving.load()                               int ns := now_serving
        if ns = my_ticket                                          if ¬XRELEASE CAS(&next_ticket,
            break                                                      ns+1, ns, RW‖)
        pause(base × (my_ticket − ns))                             now_serving.store(ns + 1)
    fence(R‖RW)
```

Figure 9.1: The Ticket lock of Figure 4.7, modified to make use of hardware lock elision.

ical sections on the same lock simultaneously, each of them invisibly incrementing and restoring next_ticket, and never changing now_serving.

9.2.3   HYBRID TM

While faster than the all-software TM implementations of Section 9.1, HTM systems seem likely, for the foreseeable future, to have limitations that will sometimes lead them to abort even in the absence of conflicts. It seems reasonable to hope that a hardware/software *hybrid* might combine (most of) the speed of the former with the generality of the latter.

Several styles of hybrid have been proposed. In some, hardware serves only to accelerate a TM system implemented primarily in software. In others, hardware implements complete support for some subset of transactions. In this latter case, the hardware and software may be designed together, or the software may be designed to accommodate a generic "best effort" HTM.

"Best effort" hybrids have the appeal of compatibility with near-term commercial HTM. If transactions abort due to conflicts, the only alternative to rewriting the application would seem to be fallback to a global lock. If transactions abort due to hardware limitations, however, fallback to software transactions would seem to be attractive.

Hardware-accelerated STM

Experimental results with a variety of STM systems suggest a baseline overhead (single-thread slowdown) of 3–10× for atomic operations. Several factors contribute to this overhead, including conflict detection, the buffering of speculative writes (undo or redo logging), validation to ensure consistency, and conflict resolution (arbitration among conflicting transactions). All of these are potentially amenable to hardware acceleration.

Saha et al. [2006b] propose to simplify conflict detection by providing hardware *mark bits* on cache lines. Set and queried by software, these bits are cleared when a cache line is invalidated— e.g., by remote access. To avoid the need to poll the bits, Spear et al. [2007] propose a general-purpose *alert-on-update* mechanism that a triggers a software handler when a marked line is accessed remotely. Minh et al. [2007] propose an alternative conflict detection mechanism based on hardware read and write *signatures* (Bloom filters).

Shriraman et al. [2007] propose to combine in-cache hardware buffering of speculative cache lines with software conflict detection and resolution; alert-on-update provides immediate notification of conflicts, eliminating the need for validation. In subsequent work, the authors add signatures and *conflict summary tables*; these support eager conflict detection in hardware, leaving software responsible only for conflict resolution, which may be lazy if desired [Shriraman et al., 2010]. As suggested by Hill et al. [2007], the "decoupling" of mechanisms for access tracking, buffering, notification, etc. serves to increase their generality: in various other combinations they can be used for such non-TM applications as debugging, fine-grain protection, memory management, and active messaging.

Hardware/Software TM Codesign

In hardware-assisted STM, atomicity remains a program-level property, built on multiple (non-atomic) hardware-level operations. To maximize performance, one would presumably prefer to implement atomicity entirely in hardware. If hardware transactions are sometimes unsuccessful for reasons other than conflicts, and if fallback to a global lock is not considered acceptable, the challenge then becomes to devise a fallback mechanism that interoperates correctly with hardware transactions.

One possible approach is to design the hardware and software together. Kumar et al. [2006] propose an HTM to complement the object-cloning DSTM of Herlihy et al. [2003b]. Baugh et al. [2008] assume the availability of fine-grain memory protection [Zhou et al., 2004], which they use in software transactions to force aborts in conflicting hardware transactions. A more common approach assumes that the hardware is given, and designs software to go with it.

Best-effort Hybrid TM

An HTM implementation is termed "best effort" if it makes no guarantees of completion, even in the absence of conflicts, and makes no assumptions about the nature of software transactions that might be running concurrently. All of the commercial HTM systems discussed in Section 9.2.1—with the exception of constrained transactions in z TM—fit this characterization.

If "spurious" aborts are common enough to make fallback to a global lock unattractive, one is faced with the question of how to make an STM fallback interoperate with HTM—in particular, how to notice when transactions of different kinds conflict. One side of the interaction is straightforward: if a software transaction writes—either eagerly or at commit time—a location that has been read or written by a hardware transaction, the hardware will abort. The other side is harder: if, say, a software transaction reads location X, a hardware transaction commits changes to both X and Y, and then the software transaction reads Y, how are we to know that the software transaction has seen inconsistent versions, and needs to abort?

Perhaps the most straightforward option, suggested by Damron et al. [2006], is to add extra instructions to the code of hardware transactions, so they update the software metadata of any locations they write. Software transactions can then inspect this metadata to validate their consistency, just as they would in an all-software system. Unfortunately, metadata updates can significantly slow the HTM code path. Vallejo et al. [2011] show how to move much of the instrumentation inside if (hw_txn) conditions, but the condition tests themselves still incur nontrivial overhead. For object-oriented languages, Tabba et al. [2009] show how instrumented hardware transactions can safely make in-place updates to objects that are cloned by software transactions.

To eliminate the need for instrumentation on the hardware path, Lev et al. [2007] suggest never running hardware and software transactions concurrently. Instead, they switch between hardware and software *phases* on a global basis. Performance can be excellent, but also somewhat brittle: unless software phases are rare, global phase changes can introduce significant delays.

Arguably the most appealing approach to best-effort hybrid TM is to employ an STM algorithm that can detect the execution of concurrent hardware transactions without the need to instrument HTM loads and stores. Dalessandro et al. [2011] achieve this goal by using NOrec (Section 9.1.4) on the software path, to leverage value-based validation. Significantly, the scalability limitation imposed by NOrec's serial write-back is mitigated in the hybrid version by counting on most transactions to finish in hardware—STM is only a fallback.

9.3   CHALLENGES

To serve its original purpose—to facilitate the construction of small, self-contained concurrent data structures—TM need not be exposed at the programming language level. Much as expert programmers use CAS and other atomic hardware instructions to build library-level synchronization mechanisms and concurrent data structures, so might they use HTM to improve performance "under the hood," without directly impacting "ordinary" programmers. Much of the appeal of TM, however, is its potential to help those programmers write parallel code that is both correct and scalable. To realize this potential, TM must be integrated into language semantics and implementations. In this final section of the lecture, we discuss some of the issues involved in this integration. Note that the discussion raises more questions than it answers: as of this writing, language support for TM is still a work in progress.

9.3.1   SEMANTICS

The most basic open question for TM semantics is "what am I allowed to do inside?" Some operations—interactive I/O in particular—are incompatible with speculation. (We cannot tell the human user "please forget I asked you that.") Rollback of other operations—many system calls among them—may be so difficult as to force a nonspeculative implementation. The two most obvious strategies for such *irreversible* (*irrevocable*) operations are to (1) simply disallow them in transactions, or (2) force a transaction that performs them to become *inevitable*—i.e., guaranteed to commit. While Spear et al. [2008b] have shown that inevitability does not always necessitate mutual exclusion, it nonetheless imposes severe constraints on scalability.

Some TM implementations are nonblocking, as discussed in Section 9.1.1. Should this property ever be part of the language-level semantics? Without it, performance may be less predictable on multiprogrammed cores, and event-driven code may be subject to deadlock if handlers cannot be preempted.

In its role as a synchronization mechanism, language-level TM must be integrated into the language memory model (Section 3.4). Some researchers have argued that since locks already form the basis of many memory models, the behavior of transactions should be defined in terms of implicit locking [Menon et al., 2008]. The C++ standards committee, which is currently considering TM language extensions [Adl-Tabatabai et al., 2012], is likely to adopt semantics in which transactions are co-equal with locks and atomic variable access—all three kinds of operations will contribute to a program's synchronization order. Given, however, that TM is often promoted as a higher-level, more intuitive alternative to locks, there is something conceptually unsatisfying about defining transactional behavior *in terms of* (or even in concert with) the thing it is supposed to replace. Clearly any language that allows transactions and locks in the same program must explain how the two interact. Considering that locks are typically implemented using lower-level atomic operations like CAS, a potentially appealing approach is to turn the tables, as it were, and define locks in terms of atomic blocks [Dalessandro et al., 2010b]. In the framework of Section 3.4, a global total order on transactions provides a trivial synchronization order, which combines with program order to yield the overall notion of happens-before. In the resulting framework, one can easily show that a data-race-free program is *transactionally sequentially consistent*: all memory accesses appear to happen in a global total order that is consistent with program order in each thread, and that keeps the accesses of any given transaction contiguous.

Some challenges of language integration are more pedestrian: If there are limits on the operations allowed inside transactions, should these be enforced at compile time or at run time? If the former, how do we tell whether it is safe to call a subroutine that is defined in a different compilation unit? Must the subroutine interface explicitly indicate whether it is "transaction safe"?

Strong and Weak Isolation

As we noted at the beginning of Section 9.2, most HTM systems are *strongly atomic*—their transactions serialize not only with other transactions, but also with individual loads and stores. Some researchers have argued [Abadi et al., 2009, Baugh et al., 2008, Blundell et al., 2005, Schneider et al., 2008, Shpeisman et al., 2007] that language-level TM should guarantee strong atomicity as well, though this is difficult to implement in software. The difference between strong and weak atomicity, however, can be seen only in programs with data races—races between transactional and non-transactional accesses in particular [Dalessandro and Scott, 2009]. If data races are considered to be bugs, then strong atomicity serves only to make the behavior of buggy programs easier to diagnose—race-free programs can never tell the difference.

Other challenges involve "surprise" interactions with other language features. Exceptions are particularly problematic. If an exception arises in a transaction and is not caught internally, what should happen when its propagation reaches the transaction boundary? Should the transaction commit or abort? If the exception represents an error, committing may not be safe: the transaction may not have restored program invariants. On the other hand, if the transaction aborts, and "never happened," how can it raise an exception? The answer may depend, at least in part, on whether speculation is part of the language semantics. Some researchers argue "no"—transactions should simply be atomic, and speculation (if any) should be an implementation detail. Other researchers argue that not only exceptions but also inevitability, explicit aborts, and condition synchronization (the latter two of which we consider in Section 9.3.2 below) are easier to understand if the programmer thinks in terms of speculation.

9.3.2   EXTENSIONS

When adding transactions to a programming language, one may want—or need—to include a variety of features not yet discussed in this chapter.

Nesting

In the chapter introduction we argued that one of the key advantages of transactions over lock-based critical sections was their *composability*. Composability requires that we allow transactions to nest. The simplest way to do so is to "flatten" them—to *subsume* the inner transaction(s) in the outer, and allow the entire unit to commit or abort together. All current commercial HTM implementations provide subsumption nesting, generally with some maximum limit on depth. Several STM systems do likewise.

For performance reasons, it may sometimes be desirable to allow an inner transaction to abort and retry while retaining the work that has been done so far in the outer transaction. This option, known as "true" or *closed* nesting, will also be required in any system that allows a transaction to abort and *not* retry. We have already considered such a possibility for exceptions that escape transactions. It will also arise in any language that provides the programmer with an explicit abort command [Harris et al., 2005].

For the sake of both performance and generality, it may also be desirable to allow concurrency within transactions—e.g., to employ multiple threads in a computationally demanding operation, and commit their results atomically [Agrawal et al., 2008].

In some cases it may even be desirable to allow an inner transaction to commit when the surrounding transaction aborts [Moss and Hosking, 2006, Ni et al., 2007]. This sort of *open* nesting may violate serializability, and must be used with care. Possible applications include the preservation of semantically neutral but performance-advantageous operations like garbage collection, memoization, and rebalancing; the collection of debugging or performance information; and the construction of "boosted" abstractions (Section 9.1.3).

Condition Synchronization

Like lock-based critical sections, transactions sometimes depend on preconditions, which may or may not hold. In Chapter 5 we considered a variety of mechanisms whereby a thread could wait for a precondition in a critical section. But a transaction cannot wait: because it is isolated, changes to the state of the world, made by other threads, will not be visible to it.

There is an analogy here to nonblocking operations, which cannot wait and still be non-blocking. The analogy suggests a potential solution: insist that transactions be total—that their preconditions always be true—but allow them to commit "reservation" notices in the style of dual data structures (Section 8.7). If, say, a dequeue operation on a transactional queue finds no data to remove, it can enqueue a reservation atomically instead, and return an indication that it has done so. The surrounding code can then wait for the reservation to be satisfied in normal, nontransactional code.

A second alternative, suggested by Smaragdakis et al. [2007], is to suspend ("punctuate") a transaction at a conditional wait, and to make the sections of the transaction before and after the wait individually (but not jointly) atomic. This alternative requires, of course, that any invariants maintained by the transaction be true at the punctuation point. If a wait may be nested inside called routines, the fact that they may wait probably needs to be an explicit part of their interface.

Perhaps the most appealing approach to transactional condition synchronization is the retry primitive of Harris et al. [2005]. When executed by a transaction, it indicates that the current operation cannot proceed, and should abort, to be retried at some future time. Exactly *when* to retry is a question reminiscent of conditional critical regions (Section 7.4.1). There is a particularly elegant answer for STM: The transaction is sure to behave the same the next time around if it reads the same values from memory. Therefore, it should become a visible reader of every location in its read set, and wait for one of those locations to be modified by another transaction. (Modification by nontransactional code would imply the existence of a data race.) The wakeup mechanism for condition synchronization is then essentially the same as the abort mechanism for visible readers, and can share the same implementation.

Other Uses of Speculation

Transactional memory is not the only potential use of speculation. Given a speculation and roll-back mechanism implemented for TM, we may consider using it for other things as well. Possibilities include

- try blocks that roll back to their original state instead of stopping where they are when an exception arises. Shinnar et al. [2004] refer to such blocks as "try-all."

- automatic or semi-automatic (programmer-hint-driven) parallelization of semantically sequential loops [Berger et al., 2009, Ding et al., 2007]. In such a loop, each iteration is essentially a transaction, but conflicts are always resolved in favor of the earlier iteration, and no iteration is permitted to commit until all its predecessors have done so. Blue Gene/Q provides explicit hardware support for such *ordered speculation*. It can also be implemented on HTM systems (e.g., Power TM) that allow nontransactional loads inside transactions.

- *safe futures* (Section 7.4.2) [Welc et al., 2005], which abort and roll back their continuation in the event of conflict, and can thus be used in languages with side effects, without changing program semantics. Like ordered speculation, safe futures can be implemented with an HTM system that supports nontransactional loads inside transactions.

9.3.3   IMPLEMENTATION

The discussion of STM in Section 9.1 conveys some sense of the breadth of possible implementation strategies. It is far from comprehensive, however. Drawing inspiration from database systems, several groups have considered *multi-version* STM systems, which increase the success rate for long-running read-only transactions by keeping old versions of modified data [Cachopo and Rito-Silva, 2006, Lu and Scott, 2012, Perelman et al., 2011, Riegel et al., 2006]. Instead of requiring that all loaded values be correct as of commit time (and then aborting every time a location in the read set is updated by another transaction), multi-version TM systems arrange for a reader to use the values that were current as of its start time, and thus to "commit in the past." To increase concurrency, it is conceivable that TM might also adopt the ability of some database systems to *forward* updates from one (still active) transaction to another, making the second dependent on the first [Ramadan et al., 2008].

The Privatization Problem

Informally, a transaction is said to *privatize* a data structure X if, prior to the transaction, X may be accessed by more than one thread, but after the transaction program logic guarantees that X is private to some particular thread. The canonical example of privatization arises with shared containers through which threads pass objects to one another. In a program with such a container, the convention may be that once an object has been removed from the container, it "belongs" to the thread that removed it, which can safely operate on it without any synchronization. If the thread returns the object to the same or a different container at a later time, it is said to *publish* the object. Publication of most shared objects also occurs at creation time: a thread typically allocates an object and initializes it before making it visible (publishing it) to other threads. Prior to publication, no synchronization is required. Dalessandro et al. [2010b] have observed that privatization is semantically equivalent to locking—it renders a shared object temporarily private. Publication is equivalent to unlocking—it makes the private object shared again.

In their usual form, publication and privatization are race-free idioms, at least at the level of the programming model: any accesses by different threads are always ordered by an intervening transaction. Unfortunately, in many STM systems, privatization is *not* race-free at the implementation level. Races arise for two reasons, and may lead to incorrect behavior in programs that are logically correct. First, completed transactions may perform "cleanup" operations (writeback of redo or undo logs) after their serialization point. These cleanup writes may interfere with nontransactional reads in the thread that now owns the privatized data. Second, "zombie" transactions, which are doomed to abort but have not yet realized this fact, may read locations that are written nontransactionally by the thread that now owns the privatized data. The result may be an inconsistent view of memory, which can cause the zombie to display erroneous, externally visible behavior.

Early STM systems did not experience the "privatization problem" because they assumed (implicitly or explicitly) that any datum that was ever accessed by more than one thread was always accessed transactionally. One solution to the privatization problem is thus to *statically partition* data into "always private" and "sometimes shared" categories. Unfortunately, attempts to enforce this partition via the type system lead to programs in which utility routines and data structures must be "cloned" to create explicitly visible transactional and nontransactional versions [Dalessandro et al., 2007].

Absent a static partition of data, any modern STM system must be "privatization safe" to be correct. Systems that serialize cleanup—RingSTM and NOrec among them—are naturally so. Others can be made so with extra instrumentation. Marathe et al. [2008] describe and evaluate several instrumentation alternatives. They identify an adaptive strategy whose performance is stable across a wide range of workloads. Dice et al. [2010] describe an additional mechanism that can be used to reduce the cost of privatization when the number of active transactions is significantly smaller than the number of extant threads. Even so, the overheads remain significant—enough so that one must generally dismiss reported performance numbers for any prototype STM system that is not privatization safe.

Publication, it turns out, can also lead to unexpected or erroneous behavior, but only in the presence of program-level data races between transactional and nontransactional code [Menon et al., 2008]. If data races are viewed as bugs, the "publication problem" can safely be ignored.

Compilation

While many researchers once expected that TM might be successfully implemented in a library/run-time system, most now agree that it requires language integration and compiler support. Compilers can be expected to instrument transactional loads and stores; clone code paths for nontransactional, STM, and HTM execution; and insert validation where necessary to sandbox dangerous operations. They can also be expected to implement a variety of performance optimizations:

- Identify accesses that are sure to touch the same location, and elide redundant instrumentation [Harris et al., 2006].

- Identify loads and stores that are certain to access private variables, and refrain from instrumenting them [Shpeisman et al., 2007]. (This task is rather tricky: if an access may touch either a shared datum or a private datum, then it must be instrumented. If the system uses redo logs, then any *other* accesses to the same private datum must also be instrumented, to ensure that a transaction always sees its own writes.)

- For strings of successive accesses, infer the minimum number of synchronizing instructions required to maintain sequential consistency [Spear et al., 2009b].

More exotic optimizations may also be possible. Olszewski et al. [2007] have proposed dynamic binary rewriting to allow arbitrary library routines to be instrumented on the fly, and called from within transactions. More ambitiously, if subroutine foo is often called inside transactions, there may be circumstances in which any subset of its arguments are known to be private, or to have already been logged. To exploit these circumstances, a compiler may choose to generate a custom clone of foo that elides instrumentation for one or more parameters.

As of this writing, compilers have been developed for transactional extensions to a variety of programming languages, including Java [Adl-Tabatabai et al., 2006, Olszewski et al., 2007], C# [Harris et al., 2006], C [Wang et al., 2007], C++ [Free Software Foundation, 2012, Intel, 2012, VELOX Project, 2011], and Haskell [HaskellWiki, 2012]. Language-level semantics are currently the most mature in Haskell, though the implementation is slow. Among more "mainstream" languages, C++ is likely to be the first to incorporate TM extensions into the language standard [Adl-Tabatabai et al., 2012].

9.3.4   DEBUGGING AND PERFORMANCE TUNING

TM also raises significant issues for debugging and performance analysis and tuning. The most obvious issue for debugging is the tension between the desire on the one hand to support debugging at the level of statements or machine instructions and, on the other hand, the need for transactions to effect their memory updates as a single indivisible operation.

Zyulkyarov et al. [2011, 2010] propose to differentiate, explicitly, between debugging an *application*, debugging the *code inside atomic blocks*, and debugging the *TM implementation*. For the first task, transactions should appear to be atomic. When single-stepping, for example, they should execute "all at once." For the second task, the debugger should ensure that threads other than the one in the transaction are quiescent, and that the values of all variables not under modification by the transaction should appear to be consistent. The third task is then akin to conventional debugging of library and runtime code.

Without new tools, programmers are likely to find it very difficult to debug transactional programs—in effect, a conventional debugger supports only the third of the TM debugging tasks. Properly supported, however, debugging of transactional programs may actually be *easier* than conventional debugging.

Herlihy and Lev [2009, 2010] propose a standard API for communication between a transactional debugger and the underlying STM system. Their work can be seen as focusing on the second of the TM debugging tasks. It requires that transactions execute in software, even if production runs employ an HTM. It also requires that the debugger differentiate, explicitly, between the state of the transaction and that of the rest of the program: if focus shifts from the transaction to that of some other thread, memory should appear to be consistent.

Unlike a lock-based critical section, a transaction never sees changes caused by other threads. A programmer who single-steps a thread through a transaction can therefore be sure that all observed changes were caused by the active thread. The STM system, moreover, can be expected to maintain extensive information of use to the debugging task, including read and write sets, both speculative and prior (consistent) versions of modified data, and conflicts between threads. In particular, if a conflict forces a thread to wait or to abort, a transactional debugger can see and report the cause.

Similar information can be expected to be of use to performance analysis tools. Conventional tools, again, are unlikely to be adequate: aborted transactions can be expected to have a major impact on performance, but isolation, if naively implemented, will leave no trace of their activity [Porter and Witchel, 2010]. Clearly some sort of information needs to "leak" out of aborted transactions. HTM designers (e.g., for Power and x86) have devoted considerable attention to the interaction of transactions with hardware performance counters. Systems software can, to a large extent, control which counters track events of committed instructions only, and which include (or also include) events performed and then elided by aborted speculation.

With traditional lock-based critical sections, performance analysis tools typically focus on acquire operations with long wait times. These identify conflicts between critical sections, which then prompt the programmer to explore more fine-grain locking alternatives. With transactions, programmers may begin, analogously, with large atomic blocks. When a performance analysis tool discovers that these often abort, it can identify the conflicts among them, prompting the programmer to search for ways to express the algorithm with smaller, more fine-grain transactions. Extensions to the programming model, such as the *early release* of Herlihy et al. [2003b] or the *elastic transactions* of Felber et al. [2009] may also allow the programmer to reduce conflicts among large transactions by explicitly evicting certain locations from the read set (e.g., if they were used only during a search phase, which can be independently verified).

The need to break transactions into smaller atomic pieces, in order to minimize conflicts, raises correctness issues: the programmer must somehow verify that decomposed operations still serialize and that the individual transactions maintain (now presumably somewhat more complex) program invariants. By continuing to use transactions instead of locks, however, the programmer can be sure (at least in the absence of condition synchronization) that the decomposition process will never introduce deadlock, and that the program will remain data-race free if shared objects are accessed only in transactions. These observations suggest that time spent diagnosing correctness bugs in lock-based programs may be replaced by time spent diagnosing performance bugs in transactional code. This change alone may prove to be the most lasting benefit of transactional synchronization.

Bibliography

Martín Abadi, Tim Harris, and Mojitaba Mehrara. Transactional memory with strong atomicity using off-the-shelf memory protection hardware. In *Proceedings of the Fourteenth ACM Symposium on Principles and Practice of Parallel Programming (PPoPP)*, pages 185–196, Raleigh, NC, February 2009. DOI: 10.1145/1504176.1504203. 166

Nagi M. Aboulenein, James R. Goodman, Stein Gjessing, and Philip J. Woest. Hardware support for synchronization in the scalable coherent interface (SCI). In *Proceedings of the Eighth International Parallel Processing Symposium (IPPS)*, pages 141–150, Cancun, Mexico, April 1994. DOI: 10.1109/IPPS.1994.288308. 25, 56

Ali-Reza Adl-Tabatabai, Brian T. Lewis, Vijay Menon, Brian R. Murphy, Bratin Saha, and Tatiana Shpeisman. Compiler and runtime support for efficient software transactional memory. In *Proceedings of the Twenty-seventh ACM SIGPLAN Conference on Programming Language Design and Implementation (PLDI)*, pages 26–37, Ottawa, ON, Canada, June 2006. DOI: 10.1145/1133255.1133985. 171

Ali-Reza Adl-Tabatabai, Tatiana Shpeisman, Justin Gottschlich, et al. *Draft Specification of Transactional Language Constructs for C++, Version 1.1*. Intel, Oracle, IBM, and Red Hat, February 2012. http://www.open-std.org/Jtc1/sc22/wg14/www/docs/n1613.pdf 47, 166, 171

Sarita V. Adve and Kourosh Gharachorloo. Shared memory consistency models: A tutorial. *Computer*, 29(12):66–76, December 1996. DOI: 10.1109/2.546611. 15, 42

Sarita V. Adve and Mark D. Hill. Weak ordering—A new definition. In *Proceedings of the Seventeenth International Symposium on Computer Architecture (ISCA)*, pages 2–14, Seattle, WA, May 1990. DOI: 10.1145/325096.325100. 46

Sarita V. Adve, Vijay S. Pai, and Parthasarathy Ranganathan. Recent advances in memory consistency models for hardware shared-memory systems. *Proceedings of the IEEE*, 87(3):445–455, 1999. DOI: 10.1109/5.747865. 15

Kunal Agrawal, Jeremy Fineman, and Jim Sukha. Nested parallelism in transactional memory. In *Proceedings of the Thirteenth ACM Symposium on Principles and Practice of Parallel Programming (PPoPP)*, pages 163–174, Salt Lake City, UT, February 2008. DOI: 10.1145/1345206.1345232. 167

Randy Allen and Ken Kennedy. *Optimizing Compilers for Modern Architectures: A Dependence-Based Approach*. Morgan Kaufmann, San Francisco, CA, 2002. 12

George S. Almasi and Allan Gottlieb. *Highly Parallel Computing*. Benjamin Cummings, Redwood City, CA, 1989. 73

Noga Alon, Amnon Barak, and Udi Manber. On disseminating information reliably without broadcasting. In *Proceedings of the International Conference on Distributed Computing Systems (ICDCS)*, pages 74–81, Berlin, Germany, September 1987. 75

AMD. *Advanced Synchronization Facility: Proposed Architectural Specification*. Advanced Micro Devices, March 2009. Publication #45432, Version 2.1. Available as amddevcentral.com/assets/45432-ASF_Spec_2.1.pdf. 160

C. Scott Ananian, Krste Asanovic Bradley C. Kuszmaul, Charles E. Leiserson, and Sean Lie. Unbounded transactional memory. In *Proceedings of the Eleventh International Symposium on High Performance Computer Architecture (HPCA)*, pages 316–327, San Francisco, CA, February 2005. DOI: 10.1109/HPCA.2005.41. 157

James Anderson and Mark Moir. Universal constructions for large objects. *IEEE Transactions on Parallel and Distributed Systems*, 10(12):1317–1332, December 1999. DOI: 10.1109/71.819952. 145

Thomas E. Anderson, Edward D. Lazowska, and Henry M. Levy. The performance of spin lock alternatives for shared-memory multiprocessors. *IEEE Transactions on Parallel and Distributed Systems*, 1(1):6–16, January 1990. DOI: 10.1109/71.80120. 54, 56

Tom E. Anderson, Brian N. Bershad, Edward D. Lazowska, and Henry M. Levy. Scheduler activations: Effective kernel support for the user-level management of parallelism. *ACM Transactions on Computer Systems*, 10(1):53–79, February 1992. DOI: 10.1145/121132.121151. 120

Jonathan Appavoo, Marc Auslander, Maria Burtico, Dilma Da Silva, Orran Krieger, Mark Mergen, Michal Ostrowski, Bryan Rosenburg, Robert W. Wisniewski, and Jimi Xenidis. Experience with K42, an open-source Linux-compatible scalable operating system kernel. *IBM Systems Journal*, 44(2):427–440, 2005. DOI: 10.1147/sj.442.0427. 59

Nimar S. Arora, Robert D. Blumofe, and C. Greg Plaxton. Thread scheduling for multiprogrammed multiprocessors. In *Proceedings of the Tenth Annual ACM Symposium on Parallel Algorithms and Architectures (SPAA)*, pages 119–129, Puerto Vallarta, Mexico, June–July 1998. DOI: 10.1145/277651.277678. 139

Hagit Attiya, Rachid Guerraoui, Danny Hendler, Petr Kuznetsov, Maged M. Michael, and Martin Vechev. Laws of order: Expensive synchronization in concurrent algorithms cannot be eliminated. In *Proceedings of the Thirty-eighth ACM Symposium on Principles of Programming Lan-

guages (POPL), pages 487–498, Austin, TX, January 2011. DOI: 10.1145/1926385.1926442.
47

Marc A. Auslander, David Joel Edelsohn, Orran Yaakov Krieger, Bryan Savoye Rosenburg,
and Robert W. Wisniewski. Enhancement to the MCS lock for increased functionality and
improved programmability. U.S. patent application number 20030200457 (abandoned),
October 2003. `http://appft1.uspto.gov/netacgi/nph-Parser?Sect1=PTO2&Sect2=`
`HITOFF&p=1&u=%2Fnetahtml%2FPTO%2Fsearch-bool.html&r=1&f=G&l=50&co1=AND&d=`
`PG01&s1=20030200457.PGNR.&OS=DN/20030200457&RS=DN/20030200457` 59

David Bacon, Joshua Bloch, Jeff Bogda, Cliff Click, Paul Haahr, Doug Lea, Tom May, Jan-
Willem Maessen, Jeremy Manson, John D. Mitchell, Kelvin Nilsen, Bill Pugh, and Emin Gun
Sirer. The 'double-checked locking is broken' declaration, 2001. `www.cs.umd.edu/~pugh/`
`java/memoryModel/DoubleCheckedLocking.html.` 68

Greg Barnes. A method for implementing lock-free shared data structures (extended abstract). In
Proceedings of the Fifth Annual ACM Symposium on Parallel Algorithms and Architectures (SPAA),
pages 261–270, Velen, Germany, June–July 1993. DOI: 10.1145/165231.165265. 145

Hans W. Barz. Implementing semaphores by binary semaphores. *ACM SIGPLAN Notices*, 18
(2):39–45, February 1983. DOI: 10.1145/948101.948103. 106

Lee Baugh, Naveen Neelakantan, and Craig Zilles. Using hardware memory protection to build
a high-performance, strongly atomic hybrid transactional memory. In *Proceedings of the Thirty-
fifth International Symposium on Computer Architecture (ISCA)*, pages 115–126, Beijing, China,
June 2008. DOI: 10.1145/1394608.1382132. 164, 166

Rudolf Bayer and Mario Schkolnick. Concurrency of operations on *B*-trees. *Acta Informatica*, 9
(1):1–21, 1977. DOI: 10.1007/BF00263762. 33

Mordechai Ben-Ari. *Principles of Concurrent and Distributed Programming*. Addison-Wesley,
2006. 41, 49

Emery Berger, Ting Yang, Tongping Liu, and Gene Novark. Grace: Safe multithreaded programming for C/C++. In *Proceedings of the Twenty-fourth Annual ACM SIGPLAN Conference
on Object-oriented Programming Systems, Languages, and Applications (OOPSLA)*, pages 81–96,
Orlando, FL, October 2009. DOI: 10.1145/1640089.1640096. 168

Mike Blasgen, Jim Gray, Mike Mitoma, and Tom Price. The convoy phenomenon. *ACM Operating Systems Review*, 13(2):20–25, April 1979. `http://dl.acm.org/citation.cfm?doid=`
`850657.850659` 119

Burton H. Bloom. Space/time trade-off in hash coding with allowable errors. *Communications
of the ACM*, 13(7):422–426, July 1970. DOI: 10.1145/362686.362692. 155

Robert D. Blumofe and Charles E. Leiserson. Scheduling multithreaded computations by work stealing. In *Proceedings of the Thirty-fifth International Symposium on Computer Foundations of Computer Science (FOCS)*, pages 356–368, Santa Fe, NM, November 1994. http://doi.ieeecomputersociety.org/10.1109/SFCS.1994.365680 139

Robert D. Blumofe, Christopher F. Joerg, Bradley C. Kuszmaul, Charles E. Leiserson, Keith H. Randall, and Yuli Zhou. Cilk: An efficient multithreaded runtime system. In *Proceedings of the Fifth ACM Symposium on Principles and Practice of Parallel Programming (PPoPP)*, pages 207–216, Santa Barbara, CA, July 1995. DOI: 10.1145/209936.209958. 116, 139

Colin Blundell, E Christopher Lewis, and Milo M. K. Martin. Deconstructing transactional semantics: The subtleties of atomicity. In *Fourth Annual Workshop on Duplicating, Deconstructing, and Debunking*, Madison, WI, June 2005. In conjunction with ISCA 2005. http://pharm.ece.wisc.edu/wddd/2005/papers/WDDD05_blundell.pdf 156, 166

Colin Blundell, Joe Devietti, E Christopher Lewis, and Milo M. K. Martin. Making the fast case common and the uncommon case simple in unbounded transactional memory. In *Proceedings of the Thirty-fourth International Symposium on Computer Architecture (ISCA)*, pages 24–34, San Diego, CA, June 2007. DOI: 10.1145/1273440.1250667. 158

Jayaram Bobba, Neelam Goyal, Mark D. Hill, Michael M. Swift, and David A. Wood. TokenTM: Efficient execution of large transactions with hardware transactional memory. In *Proceedings of the Thirty-fifth International Symposium on Computer Architecture (ISCA)*, pages 127–138, Beijing, China, June 2008. DOI: 10.1145/1394608.1382133. 157, 159

Hans-J. Boehm. Can seqlocks get along with programming language memory models? In *Proceedings of the ACM SIGPLAN Workshop on Memory Systems Performance and Correctness*, pages 12–20, Beijing, China, June 2012. DOI: 10.1145/2247684.2247688. 96

Hans-J. Boehm and Sarita V. Adve. Foundations of the C++ concurrency memory model. In *Proceedings of the Twenty-ninth ACM SIGPLAN Conference on Programming Language Design and Implementation (PLDI)*, pages 68–78, Tucson, AZ, June 2008. DOI: 10.1145/1379022.1375591. 15, 42, 46

Björn B. Brandenburg and James H. Anderson. Spin-based reader-writer synchronization for multiprocessor real-time systems. *Real-Time Systems*, 46(1):25–87, September 2010. DOI: 10.1007/s11241-010-9097-2. 89, 94

Per Brinch Hansen. *Operating System Principles*. Prentice-Hall, Englewood Cliffs, NJ, 1973. 108, 114

Per Brinch Hansen. The design of Edison. *Software—Practice and Experience*, 11(4):363–396, April 1981. DOI: 10.1002/spe.4380110404. 114

Per Brinch Hansen. The programming language Concurrent Pascal. *IEEE Transactions on Software Engineering*, SE–1(2):199–207, June 1975. DOI: 10.1109/TSE.1975.6312840. 109

Eugene D. Brooks III. The butterfly barrier. *International Journal of Parallel Programming*, 15(4): 295–307, August 1986. DOI: 10.1007/BF01407877. 75

Paul J. Brown and Ronald M. Smith. Shared data controlled by a plurality of users. U. S. patent number 3,886,525, May 1975. Filed June 1973. DRT 21

Jehoshua Bruck, Danny Dolev, Ching-Tien Ho, Marcel-Cătălin Roşu, and Ray Strong. Efficient message passing interface (MPI) for parallel computing on clusters of workstations. In *Proceedings of the Seventh Annual ACM Symposium on Parallel Algorithms and Architectures (SPAA)*, pages 64–73, Santa Barbara, CA, July 1995. DOI: 10.1145/215399.215421. 118

Sebastian Burckhardt, Rajeev Alur, and Milo M. K. Martin. CheckFence: Checking consistency of concurrent data types on relaxed memory models. In *Proceedings of the Twenty-eighth ACM SIGPLAN Conference on Programming Language Design and Implementation (PLDI)*, pages 12–21, San Diego, CA, June 2007. DOI: 10.1145/1273442.1250737. 19

James E. Burns and Nancy A. Lynch. Mutual exclusion using indivisible reads and writes. In *Proceedings of the Eighteenth Annual Allerton Conference on Communication, Control, and Computing*, pages 833–842, Monticello, IL, October 1980. A revised version of this paper was published as "Bounds on Shared memory for Mutual Exclusion", *Information and Computation*, 107(2):171–184, December 1993. http://groups.csail.mit.edu/tds/papers/Lynch/allertonconf.pdf 50

João Cachopo and António Rito-Silva. Versioned boxes as the basis for memory transactions. *Science of Computer Programming*, 63(2):172–185, December 2006. DOI: 10.1016/j.scico.2006.05.009. 169

Irina Calciu, Dave Dice, Yossi Lev, Victor Luchangco, Virendra J. Marathe, and Nir Shavit. NUMA-aware reader-writer locks. In *Proceedings of the Eighteenth ACM Symposium on Principles and Practice of Parallel Programming (PPoPP)*, pages 157–166, Shenzhen, China, February 2013. DOI: 10.1145/2442516.2442532. 88

Luis Ceze, James Tuck, Călin Caşcaval, and Josep Torrellas. Bulk disambiguation of speculative threads in multiprocessors. In *Proceedings of the Thirty-third International Symposium on Computer Architecture (ISCA)*, pages 227–238, Boston, MA, June 2006. DOI: 10.1145/1150019.1136506. 158, 159

Rohit Chandra, Ramesh Menon, Leo Dagum, David Kohr, Dror Maydan, and Jeff McDonald. *Parallel Programming in OpenMP*. Morgan Kaufmann, San Francisco, CA, 2001. http://dl.acm.org/citation.cfm?doid=35037.42270 117

Albert Chang and Mark Mergen. 801 storage: Architecture and programming. *ACM Transactions on Computer Systems*, 6(1):28–50, February 1988. DOI: 10.1145/35037.42270. 145

Philippe Charles, Christopher Donawa, Kemal Ebcioglu, Christian Grothoff, Allan Kielstra, Christoph von Praun, Vijay Saraswat, and Vivek Sarkar. X10: An object-oriented approach to non-uniform cluster computing. In *Proceedings of the Twentieth Annual ACM SIGPLAN Conference on Object-oriented Programming Systems, Languages, and Applications (OOPSLA)*, pages 519–538, San Diego, CA, October 2005. DOI: 10.1145/1094811.1094852. 116

David Chase and Yossi Lev. Dynamic circular work-stealing deque. In *Proceedings of the Seventeenth Annual ACM Symposium on Parallelism in Algorithms and Architectures (SPAA)*, pages 21–28, Las Vegas, NV, July 2005. DOI: 10.1145/1073970.1073974. 140

Dan Chazan and Willard L. Miranker. Chaotic relaxation. *Linear Algebra and its Applications*, 2 (2):199–222, April 1969. DOI: 10.1016/0024-3795(69)90028-7. 46

Weihaw Chuang, Satish Narayanasamy, Ganesh Venkatesh, Jack Sampson, Michael Van Biesbrouck, Gilles Pokam, Brad Calder, and Osvaldo Colavin. Unbounded page-based transactional memory. In *Proceedings of the Twelfth International Symposium on Architectural Support for Programming Languages and Operating Systems (ASPLOS)*, pages 347–358, San Jose, CA, October 2006. DOI: 10.1145/1168918.1168901. 158, 159

JaeWoong Chung, Chi Cao Minh, Austen McDonald, Travis Skare, Hassan Chafi, Brian D. Carlstrom, Christos Kozyrakis, and Kunle Olukotun. Tradeoffs in transactional memory virtualization. In *Proceedings of the Twelfth International Symposium on Architectural Support for Programming Languages and Operating Systems (ASPLOS)*, pages 371–381, San Jose, CA, October 2006. DOI: 10.1145/1168919.1168903. 158

Austin T. Clements, M. Frans Kaashoek, and Nickolai Zeldovich. Scalable address spaces using RCU balanced trees. In *Proceedings of the Seventeenth International Symposium on Architectural Support for Programming Languages and Operating Systems (ASPLOS)*, pages 199–210, London, United Kingdom, March 2012. DOI: 10.1145/2189750.2150998. 99, 100

Cliff Click Jr. And now some hardware transactional memory comments. Author's Blog, Azul Systems, February 2009. www.azulsystems.com/blog/cliff/2009-02-25-and-now-some-hardware-transactional-memory-comments. 26, 156, 158, 161

Edward G. Coffman, Jr., Michael J. Elphick, and Arie Shoshani. System deadlocks. *Computing Surveys*, 3(2):67–78, June 1971. DOI: 10.1145/356586.356588. 28

Pierre-Jacques Courtois, F. Heymans, and David L. Parnas. Concurrent control with 'readers' and 'writers'. *Communications of the ACM*, 14(10):667–668, October 1971. DOI: 10.1145/362759.362813. 87, 88

Travis S. Craig. Building FIFO and priority-queueing spin locks from atomic swap. Technical Report TR 93-02-02, University of Washington Computer Science Department, February 1993. ftp://ftp.cs.washington.edu/tr/1993/02/UW-CSE-93-02-02.pdf 56, 59, 62

David E. Culler and Jaswinder Pal Singh. *Parallel Computer Architecture: A Hardware/Software Approach*. Morgan Kaufmann, San Francisco, CA, 1998. With Anoop Gupta. 12

Luke Dalessandro and Michael L. Scott.   Sandboxing transactional memory.   In *Proceedings of the Twenty-first International Conference on Parallel Architectures and Compilation Techniques (PACT)*, pages 171–180, Minneapolis, MN, September 2012. DOI: 10.1145/2370816.2370843. 149, 153

Luke Dalessandro and Michael L. Scott.   Strong isolation is a weak idea.   In *Fourth ACM SIGPLAN Workshop on Transactional Computing (TRANSACT)*, Raleigh, NC, February 2009. ftp://ftp.cs.washington.edu/tr/1993/02/UW-CSE-93-02-02.pdf 166

Luke Dalessandro, Virendra J. Marathe, Michael F. Spear, and Michael L. Scott. Capabilities and limitations of library-based software transactional memory in C++. In *Second ACM SIGPLAN Workshop on Transactional Computing (TRANSACT)*, Portland, OR, August 2007. http://www.cs.rochester.edu/u/scott/papers/2007_TRANSACT_RSTM2.pdf 147, 170

Luke Dalessandro, Dave Dice, Michael L. Scott, Nir Shavit, and Michael F. Spear. Transactional mutex locks. In *Proceedings of the Sixteenth International Euro-Par Conference*, pages II:2–13, Ischia-Naples, Italy, August–September 2010a. DOI: 10.1007/978-3-642-15291-7_2. 97, 153

Luke Dalessandro, Michael L. Scott, and Michael F. Spear. Transactions as the foundation of a memory consistency model. In *Proceedings of the Twenth-Fourth International Symposium on Distributed Computing (DISC)*, pages 20–34, Cambridge, MA, September 2010b. DOI: 10.1007/978-3-642-15763-9_4. 47, 166, 169

Luke Dalessandro, Michael F. Spear, and Michael L. Scott. NOrec: Streamlining STM by abolishing ownership records. In *Proceedings of the Fifteenth ACM Symposium on Principles and Practice of Parallel Programming (PPoPP)*, pages 67–78, Bangalore, India, January 2010c. DOI: 10.1145/1693453.1693464. 97, 153, 154

Luke Dalessandro, François Carouge, Sean White, Yossi Lev, Mark Moir, Michael L. Scott, and Michael F. Spear. Hybrid NOrec: A case study in the effectiveness of best effort hardware transactional memory. In *Proceedings of the Sixteenth International Symposium on Architectural Support for Programming Languages and Operating Systems (ASPLOS)*, pages 39–52, Newport Beach, CA, March 2011. DOI: 10.1145/1950365.1950373. 165

Peter Damron, Alexandra Fedorova, Yossi Lev, Victor Luchangco, Mark Moir, and Dan Nussbaum. Hybrid transactional memory. In *Proceedings of the Twelfth International Symposium on Architectural Support for Programming Languages and Operating Systems (ASPLOS)*, pages 336–346, San Jose, CA, October 2006. DOI: 10.1145/1168919.1168900. 165

Mathieu Desnoyers, Paul E. McKenney, Alan S. Stern, Michel R. Dagenais, and Jonathan Walpole. User-level implementations of read-copy update. *IEEE Transactions on Parallel and Distributed Systems*, 23(2):375–382, February 2012. DOI: 10.1109/TPDS.2011.159. 97, 98

Dave Dice, Hui Huang, and Mingyao Yang.   Asymmetric Dekker synchronization. Lead author's blog, Oracle Corp., July 2001.   `blogs.oracle.com/dave/resource/` `Asymmetric-Dekker-Synchronization.txt`. 69, 70

Dave Dice, Mark Moir, and William N. Scherer III. Quickly reacquirable locks. Technical Report, Sun Microsystems Laboratories, 2003. Subject of U.S. Patent #7,814,488. 70

Dave Dice, Ori Shalev, and Nir Shavit. Transactional locking II. In *Proceedings of the Twentieth International Symposium on Distributed Computing (DISC)*, pages 194–208, Stockholm, Sweden, September 2006. DOI: 10.1007/11864219_14. 148, 149, 154

Dave Dice, Yossi Lev, Mark Moir, and Daniel Nussbaum.   Early experience with a commercial hardware transactional memory implementation.   In *Proceedings of the Fourteenth International Symposium on Architectural Support for Programming Languages and Operating Systems (ASPLOS)*, pages 157–168, Washington, DC, March 2009.   Expanded version available as SMLI TR-2009-180, Sun Microsystems Laboratories, October 2009. DOI: 10.1145/1508244.1508263. 26, 156, 157, 159

Dave Dice, Alexander Matveev, and Nir Shavit. Implicit privatization using private transactions. In *Fifth ACM SIGPLAN Workshop on Transactional Computing (TRANSACT)*, Paris, France, April 2010. `http://people.csail.mit.edu/shanir/publications/Implicit%` `20Privatization.pdf`   or   `http://people.cs.umass.edu/~moss/transact-2010/` `public-papers/03.pdf` 170

Dave Dice, Yossi Lev, Yujie Liu, Victor Luchangco, and Mark Moir. Using hardware transactional memory to correct and simplify a readers-writer lock algorithm. In *Proceedings of the Eighteenth ACM Symposium on Principles and Practice of Parallel Programming (PPoPP)*, pages 261–270, Shenzhen, China, February 2013. DOI: 10.1145/2442516.2442542. 89, 92, 94

David Dice. Inverted schedctl usage in the JVM. Author's Blog, Oracle Corp., June 2011. `blogs.oracle.com/dave/entry/inverted_schedctl_usage_in_the`. 120

David Dice, Virendra J. Marathe, and Nir Shavit. Lock cohorting: A general technique for designing NUMA locks. In *Proceedings of the Seventeenth ACM Symposium on Principles and Practice of Parallel Programming (PPoPP)*, pages 247–256, New Orleans, LA, February 2012. DOI: 10.1145/2370036.2145848. 67

Stephan Diestelhorst, Martin Pohlack, Michael Hohmuth, Dave Christie, Jae-Woong Chung, and Luke Yen. Implementing AMD's Advanced Synchronization Facility in an out-of-order x86 core. In *Fifth ACM SIGPLAN Workshop on Transactional Computing (TRANS-ACT)*, Paris, France, April 2010. http://people.cs.umass.edu/~moss/transact-2010/public-papers/14.pdf 160

Edsger W. Dijkstra. Een algorithme ter voorkoming van de dodelijke omarming. Technical Report EWD-108, IBM T. J. Watson Research Center, early 1960s. In Dutch. Circulated privately. http://www.cs.utexas.edu/~EWD/ewd01xx/EWD108.PDF 29

Edsger W. Dijkstra. The mathematics behind the banker's algorithm. In *Selected Writings on Computing: A Personal Perspective*, pages 308–312. Springer-Verlag, 1982. DOI: 10.1007/978-1-4612-5695-3_54. 29

Edsger W. Dijkstra. Solution of a problem in concurrent programming control. *Communications of the ACM*, 8(9):569, September 1965. DOI: 10.1145/365559.365617. 2

Edsger W. Dijkstra. The structure of the 'THE' multiprogramming system. *Communications of the ACM*, 11(5):341–346, May 1968a. DOI: 10.1145/363095.363143. 2, 29

Edsger W. Dijkstra. Cooperating sequential processes. In F. Genuys, editor, *Programming Languages*, pages 43–112. Academic Press, London, United Kingdom, 1968b. Originally EWD-123, Technological University of Eindhoven, 1965. 2, 49, 105

Edsger W. Dijkstra. Hierarchical ordering of sequential processes. In Charles Antony Richard Hoare and Ronald H. Perrott, editors, *Operating Systems Techniques*, A.P.I.C. Studies in Data Processing #9, pages 72–93. Academic Press, London, England, 1972. Also *Acta Informatica*, 1(8):115–138, 1971. 108

Chen Ding, Xipeng Shen, Kirk Kelsey, Chris Tice, Ruke Huang, and Chengliang Zhang. Software behavior oriented parallelization. In *Proceedings of the Twenty-eighth ACM SIGPLAN Conference on Programming Language Design and Implementation (PLDI)*, pages 223–234, San Diego, CA, June 2007. DOI: 10.1145/1273442.1250760. 153, 168

Aleksandar Dragojević, Rachid Guerraoui, and Michał Kapałka. Stretching transactional memory. In *Proceedings of the Thirtieth ACM SIGPLAN Conference on Programming Language Design and Implementation (PLDI)*, pages 155–165, Dublin, Ireland, June 2009. DOI: 10.1145/1543135.1542494. 148, 154

Joe Duffy. Windows keyed events, critical sections, and new Vista synchronization features. Author's Blog, November 2006. www.bluebytesoftware.com/blog/2006/11/29/WindowsKeyedEventsCriticalSectionsAndNewVistaSynchronizationFeatures.aspx. 121

Jan Edler, Jim Lipkis, and Edith Schonberg. Process management for highly parallel UNIX systems. In *Proceedings of the Usenix Workshop on Unix and Supercomputers*, pages 1–17, Pittsburgh, PA, September 1988. Also available as Ultracomputer Note #136, Courant Institute of Mathematical Sciences, New York University, April 1988. http://citeseerx.ist.psu.edu/viewdoc/summary?doi=10.1.1.45.4602 120

Faith Ellen, Yossi Lev, Victor Luchangco, and Mark Moir. SNZI: Scalable NonZero Indicators. In *Proceedings of the Twenty-sixth ACM Symposium on Principles of Distributed Computing (PODC)*, pages 13–22, Portland, OR, August 2007. DOI: 10.1145/1281100.1281106. 152

Kapali P. Eswaran, Jim Gray, Raymond A. Lorie, and Irving L. Traiger. The notions of consistency and predicate locks in a database system. *Communications of the ACM*, 19(11):624–633, November 1976. DOI: 10.1145/360363.360369. 35, 146

Pascal Felber, Torvald Riegel, and Christof Fetzer. Dynamic performance tuning of word-based software transactional memory. In *Proceedings of the Thirteenth ACM Symposium on Principles and Practice of Parallel Programming (PPoPP)*, pages 237–246, Salt Lake City, UT, February 2008. DOI: 10.1145/1345206.1345241. 154

Pascal Felber, Vincent Gramoli, and Rachid Guerraoui. Elastic transactions. In *Proceedings of the Twenth-third International Symposium on Distributed Computing (DISC)*, pages 93–107, Elche/Elx, Spain, September 2009. DOI: 10.1007/978-3-642-04355-0_12. 172

Michael J. Fischer, Nancy A. Lynch, James E. Burns, and Allan Borodin. Resource allocation with immunity to limited process failure. In *Proceedings of the Twentieth International Symposium on Computer Foundations of Computer Science (FOCS)*, pages 234–254, San Juan, Puerto Rico, October 1979. DOI: 10.1109/SFCS.1979.37. 52, 55

Michael J. Fischer, Nancy A. Lynch, and Michael S. Paterson. Impossibility of distributed consensus with one faulty process. *Journal of the ACM*, 32(2):374–382, April 1985. DOI: 10.1145/3149.214121. 41

Nissim Francez. *Fairness*. Springer-Verlag, 1986. DOI: 10.1007/978-1-4612-4886-6. 41

Hubertus Franke and Rusty Russell. Fuss, futexes and furwocks: Fast userlevel locking in Linux. In *Proceedings of the Ottawa Linux Symposium*, pages 479–495, Ottawa, ON, Canada, July 2002. https://www.kernel.org/doc/ols/2002/ols2002-pages-479-495.pdf 119

Keir Fraser. *Practical Lock-Freedom*. PhD thesis, King's College, University of Cambridge, September 2003. Published as University of Cambridge Computer Laboratory technical report #579, February 2004. www.cl.cam.ac.uk/techreports/UCAM-CL-TR-579.pdf. 135, 148

Keir Fraser and Tim Harris. Concurrent programming without locks. *ACM Transactions on Computer Systems*, 25(2):article 5, May 2007. DOI: 10.1145/1233307.1233309. 146, 148

Free Software Foundation. Transactional memory in GCC, February 2012. gcc.gnu.org/wiki/TransactionalMemory. 171

Matteo Frigo, Charles E. Leiserson, and Keith H. Randall. The implementation of the Cilk-5 multithreaded language. In *Proceedings of the Nineteenth ACM SIGPLAN Conference on Programming Language Design and Implementation (PLDI)*, pages 212–223, Montreal, PQ, Canada, June 1998. DOI: 10.1145/277652.277725. 116, 139

David Gifford, Alfred Spector, Andris Padegs, and Richard Case. Case study: IBM's System/360–370 architecture. *Communications of the ACM*, 30(4):291–307, April 1987. DOI: 10.1145/32232.32233. 21

Brian Goetz, Tim Peierls, Joshua Bloch, Joseph Bowbeer, David Holmes, and Doug Lea. *Java Concurrency in Practice*. Addison-Wesley Professional, 2006. 109

James R. Goodman. Using cache memory to reduce processor-memory traffic. In *Proceedings of the Tenth International Symposium on Computer Architecture (ISCA)*, pages 124–131, Stockholm, Sweden, June 1983. DOI: 10.1145/1067651.801647. 12

James R. Goodman, Mary K. Vernon, and Philip J. Woest. Efficient synchronization primitives for large-scale cache-coherent multiprocessors. In *Proceedings of the Third International Symposium on Architectural Support for Programming Languages and Operating Systems (ASPLOS)*, pages 64–75, Boston, MA, April 1989. DOI: 10.1145/70082.68188. 25, 56

Allan Gottlieb, Ralph Grishman, Clyde P. Kruskal, Kevin P. McAuliffe, Larry Rudolph, and Marc Snir. The NYU Ultracomputer: Designing an MIMD shared memory parallel computer. *IEEE Transactions on Computers*, 32(2):175–189, February 1983. DOI: 10.1109/TC.1983.1676201. 74

Gary Graunke and Shreekant Thakkar. Synchronization algorithms for shared-memory multiprocessors. *Computer*, 23(6):60–69, June 1990. DOI: 10.1109/2.55501. 56

Rachid Guerraoui and Michał Kapałka. On the correctness of transactional memory. In *Proceedings of the Thirteenth ACM Symposium on Principles and Practice of Parallel Programming (PPoPP)*, pages 175–184, Salt Lake City, UT, February 2008. DOI: 10.1145/1345206.1345233. 149

Rachid Guerraoui, Maurice Herlihy, and Bastian Pochon. Polymorphic contention management in SXM. In *Proceedings of the Nineteenth International Symposium on Distributed Computing (DISC)*, pages 303–323, Cracow, Poland, September 2005a. DOI: 10.1007/11561927_23. 156

Rachid Guerraoui, Maurice Herlihy, and Bastian Pochon. Toward a theory of transactional contention managers. In *Proceedings of the Twenty-fourth ACM Symposium on Principles of Distributed Computing (PODC)*, pages 258–264, Las Vegas, NV, July 2005b. DOI: 10.1145/1073814.1073863. 156

Rajiv Gupta. The fuzzy barrier: A mechanism for high speed synchronization of processors. In *Proceedings of the Third International Symposium on Architectural Support for Programming Languages and Operating Systems (ASPLOS)*, pages 54–63, Boston, MA, April 1989. DOI: 10.1145/70082.68187. 80

Rajiv Gupta and Charles R. Hill. A scalable implementation of barrier synchronization using an adaptive combining tree. *International Journal of Parallel Programming*, 18(3):161–180, June 1989. DOI: 10.1007/BF01407897. 82

Theo Haerder and Andreas Reuter. Principles of transaction-oriented database recovery. *ACM Computing Surveys*, 15(4):287–317, December 1983. DOI: 10.1145/289.291. 146

Robert H. Halstead, Jr. Multilisp: A language for concurrent symbolic computation. *ACM Transactions on Programming Languages and Systems*, 7(4):501–538, October 1985. DOI: 10.1145/4472.4478. 115

Lance Hammond, Vicky Wong, Mike Chen, Ben Hertzberg, Brian Carlstrom, Manohar Prabhu, Honggo Wijaya, Christos Kozyrakis, and Kunle Olukotun. Transactional memory coherence and consistency. In *Proceedings of the Thirty-first International Symposium on Computer Architecture (ISCA)*, pages 102–113, München, Germany, June 2004. DOI: 10.1145/1028176.1006711. 159

Yijie Han and Raphael A. Finkel. An optimal scheme for disseminating information. In *Proceedings of the International Conference on Parallel Processing (ICPP)*, pages II:198–203, University Park, PA, August 1988. 75

Tim Harris and Keir Fraser. Language support for lightweight transactions. In *Proceedings of the Eighteenth Annual ACM SIGPLAN Conference on Object-oriented Programming Systems, Languages, and Applications (OOPSLA)*, pages 388–402, Anaheim, CA, October 2003. DOI: 10.1145/949343.949340. 146, 148

Tim Harris, Simon Marlow, Simon Peyton Jones, and Maurice Herlihy. Composable memory transactions. In *Proceedings of the Tenth ACM Symposium on Principles and Practice of Parallel Programming (PPoPP)*, pages 48–60, Chicago, IL, June 2005. DOI: 10.1145/1065944.1065952. 167, 168

Timothy Harris, Mark Plesko, Avraham Shinnar, and David Tarditi. Optimizing memory transactions. In *Proceedings of the Twenty-seventh ACM SIGPLAN Conference on Programming Lan-*

guage Design and Implementation (PLDI), pages 14–25, Ottawa, ON, Canada, June 2006. DOI: 10.1145/1133255.1133984. 148, 170, 171

Timothy L. Harris. A pragmatic implementation of non-blocking linked-lists. In *Proceedings of the Fifteenth International Symposium on Distributed Computing (DISC)*, pages 300–314, Lisbon, Portugal, October 2001. DOI: 10.1007/3-540-45414-4_21. 129

Timothy L. Harris, James R. Larus, and Ravi Rajwar. *Transactional Memory*. Morgan & Claypool, San Francisco, CA, second edition, 2010. First edition, by Larus and Rajwar only, 2007. DOI: 10.2200/S00272ED1V01Y201006CAC011. 26, 145

HaskellWiki.   Software transactional memory, February 2012.   www.haskell.org/haskellwiki/Software_transactional_memory. 171

Bijun He, William N. Scherer III, and Michael L. Scott.   Preemption adaptivity in time-published queue-based spin locks. In *Proceedings of the Twelfth International Conference on High Performance Computing*, pages 7–18, Goa, India, December 2005. DOI: 10.1007/11602569_6. 120

Danny Hendler and Nir Shavit. Non-blocking steal-half work queues. In *Proceedings of the Twenty-first ACM Symposium on Principles of Distributed Computing (PODC)*, pages 280–289, Monterey, CA, July 2002. DOI: 10.1145/571825.571876. 140

Danny Hendler, Nir Shavit, and Lena Yerushalmi. A scalable lock-free stack algorithm. In *Proceedings of the Sixteenth Annual ACM Symposium on Parallelism in Algorithms and Architectures (SPAA)*, pages 206–215, Barcelona, Spain, June 2004. DOI: 10.1145/1007912.1007944. 85, 142

Danny Hendler, Itai Incze, Nir Shavit, and Moran Tzafrir. Scalable flat-combining based synchronous queues. In *Proceedings of the Twenth-Fourth International Symposium on Distributed Computing (DISC)*, pages 79–93, Cambridge, MA, September 2010a. DOI: 10.1007/978-3-642-15763-9_8. 85

Danny Hendler, Itai Incze, Nir Shavit, and Moran Tzafrir.   Flat combining and the synchronization-parallelism tradeoff. In *Proceedings of the Twenty-second Annual ACM Symposium on Parallelism in Algorithms and Architectures (SPAA)*, pages 355–364, Thira, Santorini, Greece, June 2010b. DOI: 10.1145/1810479.1810540. 85, 128

Debra A. Hensgen, Raphael A. Finkel, and Udi Manber. Two algorithms for barrier synchronization. *International Journal of Parallel Programming*, 17(1):1–17, February 1988. DOI: 10.1007/BF01379320. 73, 75, 76

Maurice Herlihy and Eric Koskinen. Transactional boosting: A methodology for highly-concurrent transactional objects. In *Proceedings of the Thirteenth ACM Symposium on Principles and Practice of Parallel Programming (PPoPP)*, pages 207–216, Salt Lake City, UT, February 2008. DOI: 10.1145/1345206.1345237. 151

Maurice Herlihy and Yossi Lev. tm_db: A generic debugging library for transactional programs. In *Proceedings of the Eighteenth International Conference on Parallel Architectures and Compilation Techniques (PACT)*, pages 136–145, Raleigh, NC, September 2009. DOI: 10.1109/PACT.2009.23. 171

Maurice Herlihy and J. Eliot B. Moss. Transactional memory: Architectural support for lock-free data structures. In *Proceedings of the Twentieth International Symposium on Computer Architecture (ISCA)*, pages 289–300, San Diego, CA, May 1993. DOI: 10.1109/ISCA.1993.698569. 26, 143, 145

Maurice Herlihy and Nir Shavit. *The Art of Multiprocessor Programming*. Morgan Kaufmann, San Francisco, CA, 2008. 27, 123, 135

Maurice Herlihy, Victor Luchangco, and Mark Moir. Obstruction-free synchronization: Double-ended queues as an example. In *Proceedings of the International Conference on Distributed Computing Systems (ICDCS)*, pages 522–529, Providence, RI, May 2003a. DOI: 10.1109/ICDCS.2003.1203503. 38, 135, 136, 137, 138, 139

Maurice Herlihy, Victor Luchangco, Mark Moir, and William N. Scherer III. Software transactional memory for dynamic-sized data structures. In *Proceedings of the Twenty-second ACM Symposium on Principles of Distributed Computing (PODC)*, pages 92–101, Boston, MA, July 2003b. DOI: 10.1145/872035.872048. 38, 146, 148, 164, 172

Maurice Herlihy, Victor Luchangco, Paul Martin, and Mark Moir. Nonblocking memory management support for dynamic-sized data structures. *ACM Transactions on Computer Systems*, 23(2):146–196, May 2005. DOI: 10.1145/1062247.1062249. 25

Maurice P. Herlihy. Wait-free synchronization. *ACM Transactions on Programming Languages and Systems*, 13(1):124–149, January 1991. DOI: 10.1145/114005.102808. 38, 41, 143

Maurice P. Herlihy. A methodology for implementing highly concurrent data objects. *ACM Transactions on Programming Languages and Systems*, 15(5):745–770, November 1993. DOI: 10.1145/161468.161469. 143, 145

Maurice P. Herlihy and Jeannette M. Wing. Linearizability: A correctness condition for concurrent objects. *ACM Transactions on Programming Languages and Systems*, 12(3):463–492, July 1990. DOI: 10.1145/78969.78972. 31, 33

Mark D. Hill. Multiprocessors should support simple memory-consistency models. *Computer*, 31(8):28–34, August 1998. DOI: 10.1109/2.707614. 19

Mark D. Hill, Derek Hower, Kevin E. Moore, Michael M. Swift, Haris Volos, and David A. Wood. A case for deconstructing hardware transactional memory systems. Technical Report TR 1594, Department of Computer Sciences, University of Wisconsin–Madison, June 2007. `http://research.cs.wisc.edu/techreports/2007/TR1594.pdf` 164

C. Anthony R. Hoare. Monitors: An operating systems structuring concept. *Communications of the ACM*, 17(10):549–557, October 1974. DOI: 10.1145/355620.361161. 108, 109

Shane V. Howley and Jeremy Jones. A non-blocking internal binary search tree. In *Proceedings of the Twenty-fourth Annual ACM Symposium on Parallelism in Algorithms and Architectures (SPAA)*, pages 161–171, Pittsburgh, PA, June 2012. DOI: 10.1145/2312005.2312036. 135

Wilson C. Hsieh and William E. Weihl. Scalable reader-writer locks for parallel systems. In *Proceedings of the Sixth International Parallel Processing Symposium (IPPS)*, pages 216–230, Beverly Hills, CA, March 1992. DOI: 10.1109/IPPS.1992.222989. 94

IBM. *Power ISA Transactional Memory*. IBM Corporation, December 2012. RFC02183: Transactional Memory, Version 2.07. Available at `www.power.org/documentation/power-isa-transactional-memory/`. 157, 160, 161

IBM. *System/370 Principles of Operation*. IBM Corporation, fourth edition, September 1975. Publication GA22-7000-4. Available as `bitsavers.informatik.uni-stuttgart.de/pdf/ibm/370/princOps/GA22-7000-4_370_Principles_Of_Operation_Sep75.pdf`. 21, 23, 24

IBM. *System/370 Extended Architecture, Principles of Operation*. IBM Corporation, 1983. Publication SA22-7085. 23, 24

Jean Ichbiah, John G. P. Barnes, Robert J. Firth, and Mike Woodger. *Rationale for the Design of the Ada Programming Language*. Cambridge University Press, Cambridge, England, 1991. 47

Intel. C++ STM compiler, prototype edition, January 2012. `software.intel.com/en-us/articles/intel-c-stm-compiler-prototype-edition`. 171

Intel. *Intel Architecture Instruction Set Extensions Programming Reference*. Intel Corporation, February 2012. Publication #319433-012. Available as `software.intel.com/file/41417`. 157, 161

Intel. *Intel 64 and IA-32 Architectures Software Developer's Manual*. Intel Corporation, December 2011. Order number 325462-041US. `http://www.intel.com/content/dam/www/public/us/en/documents/manuals/64-ia-32-architectures-software-developer-manual-325462.pdf` 20

Intermetrics. *Ada 95 Rationale*. Intermetrics, Inc., Cambridge, MA, 1995. Available as `www.adahome.com/Resources/refs/rat95.html`. 114

Amos Israeli and Lihu Rappoport. Disjoint-access parallel implementations of strong shared memory primitives. In *Proceedings of the Thirteenth ACM Symposium on Principles of Distributed Computing (PODC)*, pages 151–160, Los Angeles, CA, August 1994. DOI: 10.1145/197917.198079. 145

Christian Jacobi, Timothy Slegel, and Dan Greiner. Transactional memory architecture and implementation for IBM System z. In *Proceedings of the Forty-fifth International Symposium on Microarchitecture (MICRO)*, Vancouver, BC, Canada, December 2012. DOI: 10.1109/MICRO.2012.12. 26, 157, 159, 160, 161

Prasad Jayanti and Srdjan Petrovic. Efficient and practical constructions of LL/SC variables. In *Proceedings of the Twenty-second ACM Symposium on Principles of Distributed Computing (PODC)*, pages 285–294, Boston, MA, July 2003. DOI: 10.1145/872035.872078. 25, 124

Eric H. Jensen, Gary W. Hagensen, and Jeffrey M. Broughton. A new approach to exclusive data access in shared memory multiprocessors. Technical Report UCRL-97663, Lawrence Livermore National Laboratory, November 1987. https://e-reports-ext.llnl.gov/pdf/212157.pdf 22

F. Ryan Johnson, Radu Stoica, Anastasia Ailamaki, and Todd C. Mowry. Decoupling contention management from scheduling. In *Proceedings of the Fifteenth International Symposium on Architectural Support for Programming Languages and Operating Systems (ASPLOS)*, pages 117–128, Pittsburgh, PA, March 2010. DOI: 10.1145/1736020.1736035. 119

JRuby. Community website. jruby.org/. 47

Jython. Project website. jython.org/. 47

Anna R. Karlin, Kai Li, Mark S. Manasse, and Susan Owicki. Empirical studies of competitive spinning for a shared-memory multiprocessor. In *Proceedings of the Thirteenth ACM Symposium on Operating Systems Principles (SOSP)*, pages 41–55, Pacific Grove, CA, October 1991. DOI: 10.1145/121132.286599. 118

Joep L. W. Kessels. An alternative to event queues for synchronization in monitors. *Communications of the ACM*, 20(7):500–503, July 1977. DOI: 10.1145/359636.359710. 115

Thomas F. Knight. An architecture for mostly functional languages. In *Proceedings of the ACM Conference on Lisp and Functional Programming (LFP)*, pages 105–112, Cambridge, MA, August 1986. DOI: 10.1145/319838.319854. 145

Alex Kogan and Erez Petrank. A methodology for creating fast wait-free data structures. In *Proceedings of the Seventeenth ACM Symposium on Principles and Practice of Parallel Programming (PPoPP)*, pages 141–150, New Orleans, LA, February 2012. DOI: 10.1145/2145816.2145835. 38

Leonidas I. Kontothanassis, Robert Wisniewski, and Michael L. Scott. Scheduler-conscious synchronization. *ACM Transactions on Computer Systems*, 15(1):3–40, February 1997. DOI: 10.1145/244764.244765. 120

Eric Koskinen and Maurice Herlihy. Dreadlocks: Efficient deadlock detection. In *Proceedings of the Twentieth Annual ACM Symposium on Parallelism in Algorithms and Architectures (SPAA)*, pages 297–303, Munich, Germany, June 2008. DOI: 10.1145/1378533.1378585. 29

Eric Koskinen, Matthew Parkinson, and Maurice Herlihy. Coarse-grained transactions. In *Proceedings of the Thirty-seventh ACM Symposium on Principles of Programming Languages (POPL)*, pages 19–30, Madrid, Spain, January 2010. DOI: 10.1145/1706299.1706304. 151

Orran Krieger, Michael Stumm, Ron Unrau, and Jonathan Hanna. A fair fast scalable reader-writer lock. In *Proceedings of the International Conference on Parallel Processing (ICPP)*, pages II:201–204, St. Charles, IL, August 1993. DOI: 10.1109/ICPP.1993.21. 89, 92

Clyde P. Kruskal, Larry Rudolph, and Marc Snir. Efficient synchronization on multiprocessors with shared memory. *ACM Transactions on Programming Languages and Systems*, 10(4):579–601, October 1988. DOI: 10.1145/48022.48024. 21, 74

KSR. *KSR1 Principles of Operation*. Kendall Square Research, Waltham, MA, 1992. 25

Sanjeev Kumar, Michael Chu, Christopher J. Hughes, Partha Kundu, and Anthony Nguyen. Hybrid transactional memory. In *Proceedings of the Eleventh ACM Symposium on Principles and Practice of Parallel Programming (PPoPP)*, pages 209–220, New York, NY, March 2006. DOI: 10.1145/1168857.1168900. 164

Michael Kuperstein, Martin Vechev, and Eran Yahav. Automatic inference of memory fences. In *Proceedings of the IEEE Conference on Formal Methods in Computer-Aided Design*, pages 111–120, Lugano, Switzerland, October 2010. DOI: 10.1145/2261417.2261438. 19

Edya Ladan-Mozes and Nir Shavit. An optimistic approach to lock-free FIFO queues. *Distributed Computing*, 20(5):323–341, February 2008. DOI: 10.1007/s00446-007-0050-0. 127

Richard E. Ladner and Michael J. Fischer. Parallel prefix computation. *Journal of the ACM*, 27(4):831–838, October 1980. DOI: 10.1145/322217.322232. 74

Christoph Lameter. Effective synchronization on Linux/NUMA systems. In *Proceedings of the Gelato Federation Meeting*, San Jose, CA, May 2005. http://www.lameter.com/gelato2005.pdf 94

Leslie Lamport. A new solution of Dijkstra's concurrent programming problem. *Communications of the ACM*, 17(8):453–455, August 1974. DOI: 10.1145/361082.361093. 50

Leslie Lamport. Time, clocks, and the ordering of events in a distributed system. *Communications of the ACM*, 21(7):558–565, July 1978. DOI: 10.1145/359545.359563. 32

Leslie Lamport. How to make a multiprocessor computer that correctly executes multiprocess programs. *IEEE Transactions on Computers*, C-28(9):690–691, September 1979. DOI: 10.1109/TC.1979.1675439. 14

Leslie Lamport. A fast mutual exclusion algorithm. *ACM Transactions on Computer Systems*, 5 (1):1–11, February 1987. DOI: 10.1145/7351.7352. 52

Butler W. Lampson and David D. Redell. Experience with processes and monitors in Mesa. *Communications of the ACM*, 23(2):105–117, February 1980. DOI: 10.1145/358818.358824. 109, 111

Doug Lea. The JSR-133 cookbook for compiler writers, March 2001. g.oswego.edu/dl/jmm/cookbook.h. 18

Craig A. Lee. Barrier synchronization over multistage interconnection networks. In *Proceedings of the Second IEEE Symposium on Parallel and Distributed Processing (SPDP)*, pages 130–133, Dallas, TX, December 1990. DOI: 10.1109/SPDP.1990.143520. 77

Daniel Lenoski, James Laudon, Kourosh Gharachorloo, Wolf-Dietrich Weber, Anoop Gupta, John Hennessy, Mark Horowitz, and Monica S. Lam. The Stanford Dash multiprocessor. *Computer*, 25(3):63–79, March 1992. DOI: 10.1109/2.121510. 20

Yossi Lev. *Debugging and Profiling of Transactional Programs*. PhD thesis, Department of Computer Science, Brown University, May 2010. http://cs.brown.edu/research/pubs/theses/phd/2010/lev.pdf 171

Yossi Lev, Mark Moir, and Dan Nussbaum. PhTM: Phased transactional memory. In *Second ACM SIGPLAN Workshop on Transactional Computing (TRANSACT)*, Portland, OR, August 2007. http://www.cs.rochester.edu/meetings/TRANSACT07/papers/lev.pdf 165

Yossi Lev, Victor Luchangco, Virendra Marathe, Mark Moir, Dan Nussbaum, and Marek Olszewski. Anatomy of a scalable software transactional memory. In *Fourth ACM SIGPLAN Workshop on Transactional Computing (TRANSACT)*, Raleigh, NC, February 2009a. http://transact09.cs.washington.edu/25_paper.pdf 152

Yossi Lev, Victor Luchangco, and Marek Olszewski. Scalable reader-writer locks. In *Proceedings of the Twenty-first Annual ACM Symposium on Parallelism in Algorithms and Architectures (SPAA)*, pages 101–110, Calgary, AB, Canada, August 2009b. DOI: 10.1145/1583991.1584020. 94

Li Lu and Michael L. Scott. Unmanaged multiversion STM. In *Seventh ACM SIGPLAN Workshop on Transactional Computing (TRANSACT)*, New Orleans, LA, February 2012. `http://www.cs.rochester.edu/u/scott/papers/2012_TRANSACT_umv.pdf` 169

Boris D. Lubachevsky. Synchronization barrier and related tools for shared memory parallel programming. In *Proceedings of the International Conference on Parallel Processing (ICPP)*, pages II:175–179, University Park, PA, August 1989. DOI: 10.1007/BF01407956. 76

Nancy Lynch. *Distributed Algorithms*. Morgan Kaufmann, San Francisco, CA, 1996. 27

Peter Magnussen, Anders Landin, and Erik Hagersten. Queue locks on cache coherent multiprocessors. In *Proceedings of the Eighth International Parallel Processing Symposium (IPPS)*, pages 165–171, Cancun, Mexico, April 1994. DOI: 10.1109/IPPS.1994.288305. 56, 59, 61, 62

Jeremy Manson, William Pugh, and Sarita V. Adve. The Java memory model. In *Proceedings of the Thirty-second ACM Symposium on Principles of Programming Languages (POPL)*, pages 378–391, Long Beach, CA, January 2005. DOI: 10.1145/1047659.1040336. 15, 42, 46

Virendra J. Marathe and Mark Moir. Toward high performance nonblocking software transactional memory. In *Proceedings of the Thirteenth ACM Symposium on Principles and Practice of Parallel Programming (PPoPP)*, pages 227–236, Salt Lake City, UT, February 2008. DOI: 10.1145/1345206.1345240. 38, 148

Virendra J. Marathe, William N. Scherer III, and Michael L. Scott. Adaptive software transactional memory. In *Proceedings of the Nineteenth International Symposium on Distributed Computing (DISC)*, pages 354–368, Cracow, Poland, September 2005. DOI: 10.1007/11561927_26. 38, 148

Virendra J. Marathe, Michael F. Spear, Christopher Heriot, Athul Acharya, David Eisenstat, William N. Scherer III, and Michael L. Scott. Lowering the overhead of software transactional memory. In *First ACM SIGPLAN Workshop on Transactional Computing (TRANSACT)*, Ottawa, ON, Canada, June 2006. `http://www.cs.rochester.edu/u/scott/papers/2012_TRANSACT_umv.pdf` 148

Virendra J. Marathe, Michael F. Spear, and Michael L. Scott. Scalable techniques for transparent privatization in software transactional memory. In *Proceedings of the International Conference on Parallel Processing (ICPP)*, pages 67–74, Portland, OR, September 2008. DOI: 10.1109/ICPP.2008.69. 170

Evangelos P. Markatos. Multiprocessor synchronization primitives with priorities. In *Proceedings of the Eighth IEEE Workshop on Real-Time Operating Systems and Software*, pages 1–7, Atlanta, GA, May 1991. 62

Brian D. Marsh, Michael L. Scott, Thomas J. LeBlanc, and Evangelos P. Markatos. First-class user-level threads. In *Proceedings of the Thirteenth ACM Symposium on Operating Systems Principles (SOSP)*, pages 110–121, Pacific Grove, CA, October 1991. DOI: 10.1145/121132.344329. 120

José F. Martínez and Josep Torrellas. Speculative synchronization: Applying thread-level speculation to explicitly parallel applications. In *Proceedings of the Tenth International Symposium on Architectural Support for Programming Languages and Operating Systems (ASPLOS)*, pages 18–29, San Jose, CA, October 2002. DOI: 10.1145/605397.605400. 146

Paul E. McKenney. *Exploiting Deferred Destruction: An Analysis of Read-Copy-Update Techniques in Operating System Kernels*. PhD thesis, OGI School of Science and Engineering, Oregon Health and Science University, July 2004. http://www.rdrop.com/~paulmck/RCU/RCUdissertation.2004.07.14e1.pdf 97

Paul E. McKenney, Jonathan Appavoo, Andi Kleen, Orran Krieger, Rusty Russel, Dipankar Sarma, and Maneesh Soni. Read-copy update. In *Proceedings of the Ottawa Linux Symposium*, pages 338–367, Ottawa, ON, Canada, July 2001. Revised version available as http://www.rdrop.com/~paulmck/RCU/rclock_OLS.2001.05.01c.pdf 25, 97, 99

Avraham A. Melkman. On-line construction of the convex hull of a simple polyline. *Information Processing Letters*, 25(1):11–12, April 1987. DOI: 10.1016/0020-0190(87)90086-X. 135

John M. Mellor-Crummey and Michael L. Scott. Scalable reader-writer synchronization for shared-memory multiprocessors. In *Proceedings of the Third ACM Symposium on Principles and Practice of Parallel Programming (PPoPP)*, pages 106–113, Williamsburg, VA, April 1991a. DOI: 10.1145/109626.109637. 89, 92

John M. Mellor-Crummey and Michael L. Scott. Algorithms for scalable synchronization on shared-memory multiprocessors. *ACM Transactions on Computer Systems*, 9(1):21–65, February 1991b. A bug fix for the code in Algorithm 4, provided by Keir Fraser, can be found at www.cs.rochester.edu/research/synchronization/pseudocode/rw.html. DOI: 10.1145/103727.103729. 55, 56, 58, 75, 77

Vijay Menon, Steven Balensiefer, Tatiana Shpeisman, Ali-Reza Adl-Tabatabai, Richard L. Hudson, Bratin Saha, and Adam Welc. Practical weak-atomicity semantics for Java STM. In *Proceedings of the Twentieth Annual ACM Symposium on Parallelism in Algorithms and Architectures (SPAA)*, pages 314–325, Munich, Germany, June 2008. DOI: 10.1145/1378533.1378588. 166, 170

Michael Merritt and Gadi Taubenfeld. Computing with infinitely many processes. In *Proceedings of the Fourteenth International Symposium on Distributed Computing (DISC)*, pages 164–178, Toledo, Spain, October 2000. DOI: 10.1007/3-540-40026-5_11. 52

Robert M. Metcalfe and David R. Boggs. Ethernet: Distributed packet switching for local computer networks. *Communications of the ACM*, 19(7):395–404, July 1976. DOI: 10.1145/360248.360253. 54

Maged M. Michael. Practical lock-free and wait-free LL/SC/VL implementations using 64-bit CAS. In *Proceedings of the Eighteenth International Symposium on Distributed Computing (DISC)*, pages 144–158, Amsterdam, The Netherlands, October 2004a. DOI: 10.1007/978-3-540-30186-8_11. 124, 136

Maged M. Michael. Personal communication, March 2013. 24

Maged M. Michael. CAS-based lock-free algorithm for shared deques. In *Proeedings of the Ninth Euro-Par Conference on Parallel Processing*, pages 651–660, Klagenfurt, Austria, August 2003. DOI: 10.1007/978-3-540-45209-6_92. 135, 136, 137

Maged M. Michael. High performance dynamic lock-free hash tables and list-based sets. In *Proceedings of the Fourteenth Annual ACM Symposium on Parallel Algorithms and Architectures (SPAA)*, pages 73–82, Winnipeg, MB, Canada, August 2002. DOI: 10.1145/564870.564881. 129, 130, 131

Maged M. Michael. Hazard pointers: Safe memory reclamation for lock-free objects. *IEEE Transactions on Parallel and Distributed Systems*, 15(8):491–504, August 2004b. DOI: 10.1109/TPDS.2004.8. 25, 129, 136

Maged M. Michael. ABA prevention using single-word instructions. Technical Report RC23089, IBM T. J. Watson Research Center, January 2004c. http://www.research.ibm.com/people/m/michael/RC23089.pdf 25

Maged M. Michael and Michael L. Scott. Nonblocking algorithms and preemption-safe locking on multiprogrammed shared memory multiprocessors. *Journal of Parallel and Distributed Computing*, 51(1):1–26, January 1998. DOI: 10.1006/jpdc.1998.1446. 38, 125

Maged M. Michael and Michael L. Scott. Simple, fast, and practical non-blocking and blocking concurrent queue algorithms. In *Proceedings of the Fifteenth ACM Symposium on Principles of Distributed Computing (PODC)*, pages 267–275, Philadelphia, PA, May 1996. DOI: 10.1145/248052.248106. 125, 127

Chi Cao Minh, Martin Trautmann, JaeWoong Chung, Austen McDonald, Nathan Bronson, Jared Casper, Christos Kozyrakis, and Kunle Olukotun. An effective hybrid transactional memory system with strong isolation guarantees. In *Proceedings of the Thirty-fourth International Symposium on Computer Architecture (ISCA)*, pages 69–80, San Diego, CA, June 2007. DOI: 10.1145/1273440.1250673. 164

Mark Moir and James H. Anderson. Wait-free algorithms for fast, long-lived renaming. *Science of Computer Programming*, 25(1):1–39, October 1995. DOI: 10.1016/0167-6423(95)00009-H. 52

Mark Moir and Nir Shavit. Concurrent data structures. In Dinesh P. Metha and Sartaj Sahni, editors, *Handbook of Data Structures and Applications*, page Chapter 47. Chapman and Hall / CRC Press, San Jose, CA, 2005. 38, 123

Mark Moir, Daniel Nussbaum, ori Shalev, and Nir Shavit. Using elimination to implement scalable and lock-free FIFO queues. In *Proceedings of the Seventeenth Annual ACM Symposium on Parallelism in Algorithms and Architectures (SPAA)*, pages 253–262, Las Vegas, NV, July 2005. DOI: 10.1145/1073970.1074013. 142

Adam Morrison and Yehuda Afek . Fast concurrent queues for x86 processors. In *Proceedings of the Eighteenth ACM Symposium on Principles and Practice of Parallel Programming (PPoPP)*, pages 103–112, Shenzhen, China, February 2013. DOI: 10.1145/2442516.2442527. 128

J. Eliot B. Moss and Antony L. Hosking. Nested transactional memory: Model and architecture sketches. *Science of Computer Programming*, 63(2):186–201, December 2006. DOI: 10.1016/j.scico.2006.05.010. 167

Steven S. Muchnick. *Advanced Compiler Design and Implementation*. Morgan Kaufmann, San Francisco, CA, 1997. 12

Robert H. B. Netzer and Barton P. Miller. What are race conditions?: Some issues and formalizations. *ACM Letters on Programming Languages and Systems*, 1(1):74–88, March 1992. DOI: 10.1145/130616.130623. 45

Yang Ni, Vijay S. Menon, Ali-Reza Adl-Tabatabai, Antony L. Hosking, Richard L. Hudson, J. Eliot B. Moss, Bratin Saha, and Tatiana Shpeisman. Open nesting in software transactional memory. In *Proceedings of the Twelfth ACM Symposium on Principles and Practice of Parallel Programming (PPoPP)*, pages 68–78, San Jose, CA, March 2007. DOI: 10.1145/1229428.1229442. 167

Marek Olszewski, Jeremy Cutler, and J. Gregory Steffan. JudoSTM: A dynamic binary-rewriting approach to software transactional memory. In *Proceedings of the Sixteenth International Conference on Parallel Architectures and Compilation Techniques (PACT)*, pages 365–375, Brasov, Romania, September 2007. DOI: 10.1109/PACT.2007.4336226. 153, 171

John K. Ousterhout. Scheduling techniques for concurrent systems. In *Proceedings of the International Conference on Distributed Computing Systems (ICDCS)*, pages 22–30, Miami/Ft. Lauderdale, FL, October 1982. 118

Christos H. Papadimitriou. The serializability of concurrent database updates. *Journal of the ACM*, 26(4):631–653, October 1979. DOI: 10.1145/322154.322158. 34

Dmitri Perelman, Anton Byshevsky, Oleg Litmanovich, and Idit Keidar. SMV: Selective multi-versioning STM. In *Proceedings of the Twenty-fifth International Symposium on Distributed Computing (DISC)*, pages 125–140, Rome, Italy, September 2011. DOI: 10.1007/978-3-642-24100-0_9. 169

Gary L. Peterson. Myths about the mutual exclusion problem. *Information Processing Letters*, 12 (3):115–116, June 1981. DOI: 10.1016/0020-0190(81)90106-X. 18, 49

Gary L. Peterson and Michael J. Fischer. Economical solutions for the critical section problem in a distributed system. In *Proceedings of the Ninth ACM Symposium on the Theory of Computing (STOC)*, pages 91–97, Boulder, CO, May 1977. DOI: 10.1145/800105.803398 49

Donald E Porter and Emmett Witchel. Understanding transactional memory performance. In *Proceedings of the International Symposium on Performance Analysis of Systems and Software*, pages 97–108, White Plains, NY, March 2010. DOI: 10.1109/ISPASS.2010.5452061. 172

Sundeep Prakash, Yann-Hang Lee, and Theodore Johnson. A nonblocking algorithm for shared queues using compare-and-swap. *IEEE Transactions on Computers*, 43(5):548–559, May 1994. DOI: 10.1109/12.280802. 129

William Pugh. Skip lists: A probabilistic alternative to balanced trees. *Communications of the ACM*, 33(6):668–676, June 1990. DOI: 10.1145/78973.78977. 134

Zoran Radović and Erik Hagersten. Hierarchical backoff locks for nonuniform communication architectures. In *Proceedings of the Ninth International Symposium on High Performance Computer Architecture (HPCA)*, pages 241–252, Anaheim, CA, February 2003. DOI: 10.1109/HPCA.2003.1183542. 67

Zoran Radović and Erik Hagersten. Efficient synchronization for nonuniform communication architectures. In *Proceedings, Supercomputing 2002 (SC)*, pages 1–13, Baltimore, MD, November 2002. DOI: 10.1109/SC.2002.10038. 67

Ravi Rajwar. *Speculation-based Techniques for Lock-free Execution of Lock-based Programs*. PhD thesis, Department of Computer Sciences, University of Wisconsin–Madison, October 2002. ftp://ftp.cs.wisc.edu/galileo/papers/rajwar_thesis.ps.gz 162

Ravi Rajwar and James R. Goodman. Transactional lock-free execution of lock-based programs. In *Proceedings of the Tenth International Symposium on Architectural Support for Programming Languages and Operating Systems (ASPLOS)*, pages 5–17, San Jose, CA, October 2002. DOI: 10.1145/635506.605399. 146, 160

Ravi Rajwar and James R. Goodman.   Speculative lock elision: Enabling highly concurrent multithreaded execution. In *Proceedings of the Thirty-fourth International Symposium on Microarchitecture (MICRO)*, pages 294–305, Austin, TX, December 2001. DOI: 10.1109/MICRO.2001.991127. 146, 160

Ravi Rajwar, Maurice Herlihy, and Konrad Lai. Virtualizing transactional memory. In *Proceedings of the Thirty-second International Symposium on Computer Architecture (ISCA)*, pages 494–505, Madison, WI, June 2005. DOI: 10.1145/1080695.1070011. 158

Hany E. Ramadan, Christopher J. Rossbach, and Emmett Witchel. Dependence-aware transactional memory for increased concurrency. In *Proceedings of the Forty-first International Symposium on Microarchitecture (MICRO)*, pages 246–257, Lake Como, Italy, November 2008. DOI: 10.1109/MICRO.2008.4771795. 169

Raghavan Raman, Jisheng Zhao, Vivek Sarkar, Martin Vechev, and Eran Yahav. Scalable and precise dynamic datarace detection for structured parallelism. In *Proceedings of the Thirty-third ACM SIGPLAN Conference on Programming Language Design and Implementation (PLDI)*, pages 531–542, Beijing, China, 2012. DOI: 10.1145/2254064.2254127. 117

David P. Reed and Rajendra K. Kanodia. Synchronization with eventcounts and sequencers. *Communications of the ACM*, 22(2):115–123, February 1979. DOI: 10.1145/359060.359076. 52, 55

Torvald Riegel, Pascal Felber, and Christof Fetzer. A lazy snapshot algorithm with eager validation. In *Proceedings of the Twentieth International Symposium on Distributed Computing (DISC)*, pages 284–298, Stockholm, Sweden, September 2006. DOI: 10.1007/11864219_20. 154, 169

RSTM. Reconfigurable Software Transactional Memory website. code.google.com/p/rstm/. 150

Larry Rudolph and Zary Segall. Dynamic decentralized cache schemes for MIMD parallel processors. In *Proceedings of the Eleventh International Symposium on Computer Architecture (ISCA)*, pages 340–347, Ann Arbor, MI, June 1984. DOI: 10.1145/773453.808203. 54

Kenneth Russell and David Detlefs. Eliminating synchronization-related atomic operations with biased locking and bulk rebiasing. In *Proceedings of the Twenty-first Annual ACM SIGPLAN Conference on Object-oriented Programming Systems, Languages, and Applications (OOPSLA)*, pages 263–272, Portland, OR, October 2006. DOI: 10.1145/1167473.1167496. 69

Bratin Saha, Ali-Reza Adl-Tabatabai, adl tabatabai, Richard L. Hudson, Chi Cao Minh, and Benjamin Hertzberg. McRT-STM: A high performance software transactional memory system for a multi-core runtime. In *Proceedings of the Eleventh ACM Symposium on Principles and Practice of Parallel Programming (PPoPP)*, pages 187–197, New York, NY, March 2006a. DOI: 10.1145/1122971.1123001. 148

Bratin Saha, Ali-Reza Adl-Tabatabai, and Quinn Jacobson. Architectural support for software transactional memory. In *Proceedings of the Thirty-ninth International Symposium on Microarchitecture (MICRO)*, pages 185–196, Orlando, FL, December 2006b. DOI: 10.1109/MICRO.2006.9. 163

William N. Scherer III and Michael L. Scott. Nonblocking concurrent data structures with condition synchronization. In *Proceedings of the Eighteenth International Symposium on Distributed Computing (DISC)*, pages 174–187, Amsterdam, The Netherlands, October 2004. DOI: 10.1007/978-3-540-30186-8_13. 141

William N. Scherer III and Michael L. Scott. Advanced contention management for dynamic software transactional memory. In *Proceedings of the Twenty-fourth ACM Symposium on Principles of Distributed Computing (PODC)*, pages 240–248, Las Vegas, NV, July 2005a. DOI: 10.1145/1073814.1073861. 156

William N. Scherer III and Michael L. Scott. Randomization in STM contention management (poster). In *Proceedings of the Twenty-fourth ACM Symposium on Principles of Distributed Computing (PODC)*, Las Vegas, NV, July 2005b. www.cs.rochester.edu/u/scott/papers/2005_PODC_Rand_CM_poster_abstract.pdf. 156

William N. Scherer III, Doug Lea, and Michael L. Scott. A scalable elimination-based exchange channel. In *Workshop on Synchronization and Concurrency in Object-Oriented Languages (SCOOL)*, San Diego, CA, October 2005. In conjunction with OOPSLA 2005. http://www.cs.rochester.edu/u/scott/papers/2005_SCOOL_exchanger.pdf 142

William N. Scherer III, Doug Lea, and Michael L. Scott. Scalable synchronous queues. *Communications of the ACM*, 52(5):100–108, May 2009. DOI: 10.1145/1506409.1506431. 141

Florian T. Schneider, Vijay Menon, Tatiana Shpeisman, and Ali-Reza Adl-Tabatabai. Dynamic optimization for efficient strong atomicity. In *Proceedings of the Twenty-third Annual ACM SIGPLAN Conference on Object-oriented Programming Systems, Languages, and Applications (OOPSLA)*, pages 181–194, Nashville, TN, October 2008. DOI: 10.1145/1449764.1449779. 166

Fred B. Schneider. *On Concurrent Programming*. Springer-Verlag, 1997. DOI: 10.1007/978-1-4612-1830-2. 27, 41

Michael L. Scott. *Programming Language Pragmatics*. Morgan Kaufmann Publishers, Burlington, MA, third edition, 2009. 103

Michael L. Scott. Sequential specification of transactional memory semantics. In *First ACM SIGPLAN Workshop on Transactional Computing (TRANSACT)*, Ottawa, ON, Canada, June 2006. http://www.cs.rochester.edu/u/scott/papers/2006_TRANSACT_formal_STM.pdf 148

Michael L. Scott and John M. Mellor-Crummey. Fast, contention-free combining tree barriers for shared-memory multiprocessors. *International Journal of Parallel Programming*, 22(4):449–481, August 1994. DOI: 10.1007/BF02577741. 83

Michael L. Scott and Maged M. Michael. The topological barrier: A synchronization abstraction for regularly-structured parallel applications. Technical Report TR 605, Department of Computer Science, University or Rochester, January 1996. http://www.cs.rochester.edu/u/scott/papers/1996_TR605.pdf 117

Peter Sewell, Susmit Sarkar, Scott Owens, Francesco Zappa Nardelli, and Magnus O. Myreen. x86-TSO: A rigorous and usable programmer's model for x86 multiprocessors. *Communications of the ACM*, 53(7):89–97, July 2010. DOI: 10.1145/1785414.1785443. 20

Ori Shalev and Nir Shavit. Split-ordered lists: Lock-free extensible hash tables. *Journal of the ACM*, 53(3):379–405, May 2006. DOI: 10.1145/1147954.1147958. 132, 134

Nir Shavit and Dan Touitou. Software transactional memory. In *Proceedings of the Fourteenth ACM Symposium on Principles of Distributed Computing (PODC)*, pages 204–213, Ottawa, ON, Canada, August 1995. DOI: 10.1145/224964.224987. 143, 145

Nir Shavit and Dan Touitou. Elimination trees and the construction of pools and stacks. *Theory of Computing Systems*, 30(6):645–670, August 1997. DOI: 10.1145/215399.215419. 85

Nir Shavit and Asaph Zemach. Combining funnels: A dynamic approach to software combining. *Journal of Parallel and Distributed Computing*, 60(11):1355–1387, November 2000. DOI: 10.1006/jpdc.2000.1621. 83, 84, 85

Avraham Shinnar, David Tarditi, Mark Plesko, and Bjarne Steensgaard. Integrating support for undo with exception handling. Technical Report MSR-TR-2004-140, Microsoft Research, December 2004. http://research.microsoft.com/pubs/70125/tr-2004-140.pdf 168

Jun Shirako, David Peixotto, Vivek Sarkar, and William N. Scherer III. Phasers: A unified deadlock-free construct for collective and point-to-point synchronization. In *Proceedings of the International Conference on Supercomputing (ICS)*, pages 277–288, Island of Kos, Greece, June 2008. DOI: 10.1145/1375527.1375568. 117

Tatiana Shpeisman, Vijay Menon, Ali-Reza Adl-Tabatabai, Steven Balensiefer, Dan Grossman, Richard L. Hudson, Katherine F. Moore, and Bratin Saha. Enforcing isolation and ordering in STM. In *Proceedings of the Twenty-eighth ACM SIGPLAN Conference on Programming Language Design and Implementation (PLDI)*, pages 78–88, San Diego, CA, June 2007. DOI: 10.1145/1273442.1250744. 166, 170

Arrvindh Shriraman and Sandhya Dwarkadas. Refereeing conflicts in hardware transactional memory. In *Proceedings of the Twenty-third International Conference on Supercomputing*, pages 136–146, Yorktown Heights, NY, June 2009. DOI: 10.1145/1542275.1542299. 148

Arrvindh Shriraman, Michael F. Spear, Hemayet Hossain, Sandhya Dwarkadas, and Michael L. Scott. An integrated hardware-software approach to flexible transactional memory. In *Proceedings of the Thirty-fourth International Symposium on Computer Architecture (ISCA)*, pages 104–115, San Diego, CA, June 2007. DOI: 10.1145/1273440.1250676. 164

Arrvindh Shriraman, Sandhya Dwarkadas, and Michael L. Scott. Implementation tradeoffs in the design of flexible transactional memory support. *Journal of Parallel and Distributed Computing*, 70(10):1068–1084, October 2010. DOI: 10.1016/j.jpdc.2010.03.006. 158, 159, 164

Yannis Smaragdakis, Anthony Kay, Reimer Behrends, and Michal Young. Transactions with isolation and cooperation. In *Proceedings of the Twenty-second Annual ACM SIGPLAN Conference on Object-oriented Programming Systems, Languages, and Applications (OOPSLA)*, pages 191–210, Montréal, PQ, Canada, October 2007. DOI: 10.1145/1297027.1297042. 168

Daniel J. Sorin, Mark D. Hill, and David A. Wood. *A Primer on Memory Consistency and Cache Coherence.* Morgan & Claypool, San Francisco, CA, 2011. DOI: 10.2200/S00346ED1V01Y201104CAC016. 12, 42

Michael F. Spear, Virendra J. Marathe, William N. Scherer III, and Michael L. Scott. Conflict detection and validation strategies for software transactional memory. In *Proceedings of the Twentieth International Symposium on Distributed Computing (DISC)*, pages 179–193, Stockholm, Sweden, September 2006. DOI: 10.1007/11864219_13. 153

Michael F. Spear, Arrvindh Shriraman, Luke Dalessandro, Sandhya Dwarkadas, and Michael L. Scott. Nonblocking transactions without indirection using alert-on-update. In *Proceedings of the Nineteenth Annual ACM Symposium on Parallelism in Algorithms and Architectures (SPAA)*, pages 210–220, San Diego, CA, June 2007. DOI: 10.1145/1248377.1248414. 163

Michael F. Spear, Maged M. Michael, and Christoph von Praun. RingSTM: Scalable transactions with a single atomic instruction. In *Proceedings of the Twentieth Annual ACM Symposium on Parallelism in Algorithms and Architectures (SPAA)*, pages 275–284, Munich, Germany, June 2008a. DOI: 10.1145/1378533.1378583. 148, 154

Michael F. Spear, Michael Silverman, Luke Dalessandro, Maged M. Michael, and Michael L. Scott. Implementing and exploiting inevitability in software transactional memory. In *Proceedings of the International Conference on Parallel Processing (ICPP)*, pages 59–66, Portland, OR, September 2008b. DOI: 10.1109/ICPP.2008.55. 166

Michael F. Spear, Luke Dalessandro, Virendra J. Marathe, and Michael L. Scott. A comprehensive contention management strategy for software transactional memory. In *Proceedings of the Fourteenth ACM Symposium on Principles and Practice of Parallel Programming (PPoPP)*, pages 141–150, Raleigh, NC, February 2009a. DOI: 10.1145/1504176.1504199. 155

Michael F. Spear, Maged M. Michael, Michael L. Scott, and Peng Wu. Reducing memory ordering overheads in software transactional memory. In *Proceedings of the International Symposium on Code Generation and Optimization (CGO)*, pages 13–24, Seattle, WA, March 2009b. DOI: 10.1109/CGO.2009.30. 171

Janice M. Stone, Harold S. Stone, Philip Heidelberger, and John Turek. Multiple reservations and the Oklahoma update. *IEEE Parallel and Distributed Technology*, 1(4):58–71, November 1993. DOI: 10.1109/88.260295. 26, 145

Håkan Sundell. *Efficient and Practical Non-Blocking Data Structures*. PhD thesis, Department of Computing Science, Chalmers University of Technology, Göteborg University, 2004. www.cse.chalmers.se/~tsigas/papers/Haakan-Thesis.pdf. 123

Håkan Sundell and Philippas Tsigas. NOBLE: Non-blocking programming support via lock-free shared abstract data types. *Computer Architecture News*, 36(5):80–87, December 2008a. DOI: 10.1145/1556444.1556455. 38

Håkan Sundell and Philippas Tsigas. Lock-free deques and doubly linked lists. *Journal of Parallel and Distributed Computing*, 68(7):1008–1020, July 2008b. DOI: 10.1016/j.jpdc.2008.03.001. 128, 135

Fuad Tabba, Mark Moir, James R. Goodman, Andrew W. Hay, and Cong Wang. NZTM: Nonblocking zero-indirection transactional memory. In *Proceedings of the Twenty-first Annual ACM Symposium on Parallelism in Algorithms and Architectures (SPAA)*, pages 204–213, Calgary, AB, Canada, August 2009. DOI: 10.1145/1583991.1584048. 148, 165

Peiyi Tang and Pen-Chung Yew. Software combining algorithms for distributing hot-spot addressing. *Journal of Parallel and Distributed Computing*, 10(2):130–139, February 1990. DOI: 10.1016/0743-7315(90)90022-H. 74, 84

Gadi Taubenfeld. Shared memory synchronization. *Bulletin of the European Association for Theoretical Computer Science (BEATCS)*, (96):81–103, October 2008. http://www.faculty.idc.ac.il/gadi/MyPapers/2008T-SMsync.pdf 49

Gadi Taubenfeld. The black-white bakery algorithm. In *Proceedings of the Eighteenth International Symposium on Distributed Computing (DISC)*, pages 56–70, Amsterdam, The Netherlands, October 2004. DOI: 10.1007/978-3-540-30186-8_5. 52

Gadi Taubenfeld. *Synchronization Algorithms and Concurrent Programming*. Pearson Education–Prentice-Hall, 2006. 49

R. Kent Treiber. Systems programming: Coping with parallelism. Technical Report RJ 5118, IBM Almaden Research Center, April 1986. http://domino.research.ibm.com/library/cyberdig.nsf/papers/58319A2ED2B1078985257003004617EF/$File/rj5118.pdff 23, 24, 124

John Turek, Dennis Shasha, and Sundeep Prakash. Locking without blocking: Making lock based concurrent data structure algorithms nonblocking. In *Proceedings of the Eleventh ACM Symposium on Principles of Database Systems (PODS)*, pages 212–222, Vancouver, BC, Canada, August 1992. DOI: 10.1145/137097.137873. 145

Enrique Vallejo, Sutirtha Sanyal, Tim Harris, Fernando Vallejo, Ramón Beivide, Osman Unsal, Adrián Cristal, and Mateo Valero. Hybrid transactional memory with pessimistic concurrency control. *International Journal of Parallel Programming*, 29(3):375–396, June 2011. DOI: 10.1007/s10766-010-0158-x. 165

Nalini Vasudevan, Kedar S. Namjoshi, and Stephen A. Edwards. Simple and fast biased locks. In *Proceedings of the Nineteenth International Conference on Parallel Architectures and Compilation Techniques (PACT)*, pages 65–74, Vienna, Austria, September 2010. DOI: 10.1145/1854273.1854287. 69

VELOX Project. Dresden TM compiler (DTMC), February 2011. www.velox-project.eu/software/dtmc. 171

Jons-Tobias Wamhoff, Christof Fetzer, Pascal Felber, Etienne Rivière, and Gilles Muller. Fast-Lane: Improving performance of software transactional memory for low thread counts. In *Proceedings of the Eighteenth ACM Symposium on Principles and Practice of Parallel Programming (PPoPP)*, pages 113–122, Shenzhen, China, February 2013. DOI: 10.1145/2442516.2442528. 153

Amy Wang, Matthew Gaudet, Peng Wu, José Nelson Amaral, Martin Ohmacht, Christopher Barton, Raul Silvera, and Maged Michael. Evaluation of Blue Gene/Q hardware support for transactional memories. In *Proceedings of the Twenty-first International Conference on Parallel Architectures and Compilation Techniques (PACT)*, pages 127–136, Minneapolis, MN, September 2012. DOI: 10.1145/2370816.2370836. 26, 156

Cheng Wang, Wei-Yu Chen, Youfeng Wu, Bratin Saha, and Ali-Reza Adl-Tabatabai. Code generation and optimization for transactional memory constructs in an unmanaged language. In *Proceedings of the International Symposium on Code Generation and Optimization (CGO)*, pages 34–48, San Jose, CA, March 2007. DOI: 10.1109/CGO.2007.4. 171

William E. Weihl. Local atomicity properties: Modular concurrency control for abstract data types. *ACM Transactions on Programming Languages and Systems*, 11(2):249–282, February 1989. DOI: 10.1145/63264.63518. 33

Adam Welc, Suresh Jagannathan, and Antony L. Hosking. Safe futures for Java. In *Proceedings of the Twentieth Annual ACM SIGPLAN Conference on Object-oriented Programming Systems, Languages, and Applications (OOPSLA)*, pages 439–453, San Diego, CA, October 2005. DOI: 10.1145/1103845.1094845. 116, 169

Horst Wettstein. The problem of nested monitor calls revisited. *ACM Operating Systems Review*, 12(1):19–23, January 1978. DOI: 10.1145/850644.850645. 112

Niklaus Wirth. Modula: A language for modular multiprogramming. *Software—Practice and Experience*, 7(1):3–35, January–February 1977. DOI: 10.1002/spe.4380070102. 109

Philip J. Woest and James R. Goodman. An analysis of synchronization mechanisms in shared memory multiprocessors. In *Proceedings of the International Symposium on Shared Memory Multiprocessing (ISSMM)*, pages 152–165, Toyko, Japan, April 1991. 25

Kenneth C. Yeager. The MIPS R10000 superscalar microprocessor. *IEEE Micro*, 16(2):28–40, April 1996. DOI: 10.1109/40.491460. 19

Luke Yen, Jayaram Bobba, Michael R. Marty, Kevin E. Moore, Haris Valos, Mark D. Hill, Michael M. Swift, and David A. Wood. LogTM-SE: Decoupling hardware transactional memory from caches. In *Proceedings of the Thirteenth International Symposium on High Performance Computer Architecture (HPCA)*, pages 261–272, Phoenix, AZ, February 2007. DOI: 10.1109/HPCA.2007.346204. 157, 159

Pen-Chung Yew, Nian-Feng Tzeng, and Duncan H. Lawrie. Distributing hot-spot addressing in large-scale multiprocessors. *IEEE Transactions on Computers*, 36(4):388–395, April 1987. DOI: 10.1109/TC.1987.1676921. 74, 75

Pin Zhou, Feng Qin, Wei Liu, Yuanyuan Zhou, and Josep Torrellas. iWatcher: Efficient architectural support for software debugging. In *Proceedings of the Thirty-first International Symposium on Computer Architecture (ISCA)*, pages 224–237, München, Germany, June 2004. DOI: 10.1145/1028176.1006720. 164

Ferad Zyulkyarov. *Programming, Debugging, Profiling and Optimizing Transactional Memory Programs.* PhD thesis, Department of Computer Architecture, Polytechnic University of Catalunya (UPC), June 2011. http:http://www.feradz.com/ferad-phdthesis-20110525.pdf 171

Ferad Zyulkyarov, Tim Harris, Osman S. Unsal, Adrián Cristal, and Mateo Valero. Debugging programs that use atomic blocks and transactional memory. In *Proceedings of the Fifteenth ACM Symposium on Principles and Practice of Parallel Programming (PPoPP)*, pages 57–66, Bangalore, India, January 2010. DOI: 10.1145/1693453.1693463. 171

Author's Biography

MICHAEL L. SCOTT

Michael L. Scott is a Professor and past Chair of the Department of Computer Science at the University of Rochester. He received his Ph.D. from the University of Wisconsin–Madison in 1985. His research interests span operating systems, languages, architecture, and tools, with a particular emphasis on parallel and distributed systems. He is best known for work in synchronization algorithms and concurrent data structures, in recognition of which he shared the 2006 SIGACT/SIGOPS Edsger W. Dijkstra Prize. Other widely cited work has addressed parallel operating systems and file systems, software distributed shared memory, and energy-conscious operating systems and microarchitecture. His textbook on programming language design and implementation (*Programming Language Pragmatics*, third edition, Morgan Kaufmann, Feb. 2009) is a standard in the field. In 2003 he served as General Chair for *SOSP*; more recently he has been Program Chair for *TRANSACT'07*, *PPoPP'08*, and *ASPLOS'12*. He was named a Fellow of the ACM in 2006 and of the IEEE in 2010. In 2001 he received the University of Rochester's Robert and Pamela Goergen Award for Distinguished Achievement and Artistry in Undergraduate Teaching.

Printed in the United States
by Baker & Taylor Publisher Services